GIS for Critical Infrastructure Protection

GIS for Critical Infrastructure Protection

Robert F. Austin

David P. DiSera

Talbot J. Brooks

CRC Press
Taylor & Francis Group
Boca Raton London New York

CRC Press is an imprint of the
Taylor & Francis Group, an **informa** business

CRC Press
Taylor & Francis Group
6000 Broken Sound Parkway NW, Suite 300
Boca Raton, FL 33487-2742

First issued in paperback 2019

© 2016 by Taylor & Francis Group, LLC
CRC Press is an imprint of Taylor & Francis Group, an Informa business

No claim to original U.S. Government works

ISBN-13: 978-1-4665-9934-5 (hbk)
ISBN-13: 978-0-367-86859-8 (pbk)

Library of Congress Cataloging-in-Publication Data

Austin, Robert F.
  GIS for critical infrastructure protection / Robert F. Austin, David P. DiSera, and Talbot J. Brooks.
      pages cm
  Includes bibliographical references and index.
  ISBN 978-1-4665-9934-5 (alk. paper)
  1. Geographic information systems. 2. Infrastructure (Economics)--Security measures. 3. Emergency management. 4. First responders. I. DiSera, David P. II. Brooks, Talbot J. III. Title. IV. Title: Geographic Information Systems for critical infrastructure protection.

G70.212.A95 2016
363.34'70285--dc23   2015008596

**Visit the Taylor & Francis Web site at
http://www.taylorandfrancis.com**

**and the CRC Press Web site at
http://www.crcpress.com**

The authors dedicate this text to the brave women and men who work every day to defend this nation, its people, and its infrastructure from harm.

Contents

Preface

The purpose of this text is twofold: to provide geographic information systems professionals with an introduction to the fundamentals of critical infrastructure protection (CIP) and to provide professionals working in the field of critical infrastructure protection with an introduction to the tools and techniques of geographic information systems technology.

We begin the text with discussion of some theoretical and conceptual bases and aspects of these two bodies of knowledge—the basic knowledge that, in our opinion, is necessary for the beneficial interaction of the two fields. The first two chapters provide an introduction to the fields, while the next three chapters describe the interaction between geographic information systems (GIS) and CIP in the federal government, private sector, and local government. The next four chapters contain case studies that demonstrate how GIS and CIP are combined in the real world.

Wikipedia defines praxis as "the process by which a theory, lesson or skill is enacted, embodied or realized." There is a gap between theory and practice that is filled by praxis, which therefore also implies some element of communication or dialogue. This text is intended to assist in that dialogue and, by so doing, to help bring the theories to bear on real-world problems.

Acknowledgments

The authors, all past presidents of the Geospatial Information and Technology Association (GITA), gratefully acknowledge the efforts of GITA's volunteers and staff in guiding the evolution of geospatial analysis in the public and private sectors for four decades. We acknowledge the work of Robert Samborski and Steve Swazee in researching the history of GITA. We also acknowledge GITA's role as the sponsor of the Geospatially Enabling Community Collaboration (GECCo) program, which actively worked in cooperation with representatives of the Department of Homeland Security, the Federal Geographic Data Committee, and other federal, tribal, state, regional, and local agencies, as well as the infrastructure and emergency management communities.

Special thanks go to the hundreds of public and private participants in the GECCo events:

- City and county of Honolulu, Hawaii
- City of Denver and the Front Range, Colorado
- Western New York State, Southern Tier West Regional Planning Agency
- City of Seattle and King County, Washington
- Greater Tampa Bay area, Florida
- Greater Phoenix area, Arizona
- Greater Dallas–Ft. Worth, Texas
- Greater Twin Cities area, Minnesota
- Oakland and San Francisco Bay area, California
- City of Charlotte and Mecklenburg County, North Carolina

Chapter 3 provides a narrative of critical lessons learned through personal experience during the response to Hurricane Katrina. More than 75 volunteers answered the call for help, bringing geospatial analysis at the Mississippi Emergency Management Agency (MEMA) from an unknown tool to an indispensible technology for nearly every aspect of their work. The efforts by these volunteers often came at great personal sacrifice and were heroic. The authors thank them for answering the call to help.

Special thanks to MEMA executive director Robert Latham and Mark Sanders, MEMA geographic information systems (GIS) administrator, and Jim Steil, director of the Mississippi Automated Resource Information

System and Technical Center, for having the vision and courage to explore the potential for using spatial technologies for emergency management.

Many of the fire service examples provided throughout this work would not have been possible without the cooperation of Chief Don Oliver of Wilson, North Carolina, Fire Rescue Services, who graciously provided the data used to produce illustrations. Likewise, Colleen Heilig, training specialist for the Planning and Information Management Curriculum at the National Fire Academy, and Chiefs Dave Holmerud (retired, Solano Beach Fire Department) and Scott Avery (O'Fallon Fire Department) provided insight that guided the development of many examples as to how geospatial technologies may be applied to the fire service. We also acknowledge Lee Tedder and Lane Kimbrell for their concerted effort at helping the authors understand emergency communications systems and Rebecca Boone for her assistance with the hazard mitigation planning process.

Finally, the authors thank their families for the support and encouragement they provided. Our spouses, Michelle Austin, Natalie DiSera, and Darlene Brooks, acted as sounding boards, typists, and editors during the production of this work and frequently forgave the conference calls during vacations, the weekends spent working on the manuscript, and our general absentmindedness while focusing on this book.

Authors

Robert F. Austin is a member and the chairman of the National Geospatial Advisory Committee. Dr. Austin is a past president and past director of the Geospatial Information and Technology Association, a past member of the board of the GIS Certification Institute, and a retired senior member of the IEEE. A member of the Association of American Geographers since 1974, Dr. Austin has enjoyed three careers: as a university professor from 1977 to 1985, as a private sector scientist and consultant from 1984 to 2007, and as a manager in local government from 2007 to 2014. His studies in the field of geographic information systems began with his first college course in cartography in 1969 at the University of Michigan, where he also earned his doctorate in geography in 1977.

David P. DiSera is a vice president and chief technical officer of EMA, Inc. and a member of the National Geospatial Advisory Committee. DiSera is a past president, past director, and past research committee chair of the Geospatial Information and Technology Association (GITA). As research chair, DiSera was the Critical Infrastructure Protection Task Force leader for the Geospatially Enabling Community Collaboration Program. DiSera was a member the Federal Geographic Data Committee–Future Directions Working Committee on GIS Business Case. As a recipient of GITA's Distinguished Service Award and a two-time recipient of the Urban and Regional Information Systems Association National Leadership Award, DiSera has consulted in the field of geospatial and information technology for nearly 25 years and supported an undergraduate degree in this specialty field. DiSera has held chief information officer and chief technology officer positions, assisting utilities, local government, and the financial sector.

Talbot J. Brooks is the director of the Center for Interdisciplinary Geospatial Information Technologies at Delta State University and a member of the National Geospatial Advisory Committee. He has served in every leadership role offered by the Geospatial Information and Technology Association, is a director for the GIS Certification Institute, and chairs the Technical Users Group for the Mississippi Coordinating Council for Remote Sensing and GIS. Beyond research, policy, and teaching interests in geospatial technologies, Brooks has served as a first responder since 1987 and is an active member of the Bolivar County Volunteer Fire Department and serves as an instructor for both the Mississippi State Fire Academy and the National Fire Academy. Brooks has provided geospatial support for numerous disasters and continues to work closely with the Federal Emergency Management

Agency, Mississippi Emergency Management Agency, and the United Nations Platform for Space-Based Information for Disaster Management and Emergency Response. His work has been recognized through many awards and commendations, most notably a commendation for service passed as a joint resolution of the Mississippi legislature.

Acronyms and Abbreviations

AAR/IP	After Action Report/Improvement Plan
AM/FM	Automated Mapping/Facilities Management
ANI/ALI	Automatic Number Identification/Automatic Location Identifier
ANSI	American National Standards Institute
ASCE	American Society of Civil Engineers
BLEVE	Boiling Liquid-Expanding Vapor Explosion
BYOD	Bring Your Own Device
CAD	Computer-Aided Dispatch
CADD	Computer-Assisted Drafting and Design
CAMEO	Computer-Aided Management of Emergency Operations
CAT	Climate Assessment Tool
CATV	Community Access Television, Community Antenna Television
CERT	Community Emergency Response Team
CGA	Common Ground Alliance
CIKR	Critical Infrastructure and Key Resources
CIP	Critical Infrastructure Protection
CNES	Centre National d'Etudes Spatiales
COBOL	Common Business-Oriented Language
CODASYL	Conference on Data Systems Languages
COP	Common Operating Picture
CORS	Continuously Operating Reference System
COTS	Commercial Off-the-Shelf
COW	Cell Tower on Wheels
CRED	Centre for Research on the Epidemiology of Disasters
DaaS	Data as a Service
DBMS	Database Management System
DDL	Data Definition Language
DEM	Digital Elevation Model
DHS	Department of Homeland Security
DIME	Dual Independent Map Encoding
DIMP	Distribution Integrity Management Program
DIRT	Damage Information Reporting Tool
DMA	Designated Market Area
DOT	U.S. Department of Transportation
DRG	Digital Raster Graphic
DVD	Digital Video Disc
EMAC	Emergency Management Assistance Compact
EMPG	Emergency Management Preparedness Grant

EOC	Emergency Operations Center
EPA	Environmental Protection Agency
EPC	Emergency Preparedness Committee
ESF	Emergency Service Function
ESN	Emergency Service Number
ESS	Emergency Services Sector
ESZ	Emergency Service Zone
FBI	Federal Bureau of Investigation
FCC	Federal Communications Commission
FDEM	Florida Department of Emergency Management
FEMA	Federal Emergency Management Agency
FGDC	Federal Geographic Data Committee
FORTRAN	Formula Translation
FTP	File Transfer Protocol
GAO	U.S. Government Accountability Office
GECCo	Geospatially Enabling Community Collaboration
GeoCONOPS	Geospatial Concept of Operations
GIS	Geographic Information Systems
GITA	Geospatial Information and Technology Association
GMO	Geospatial Management Office
GNSS	Global Navigation Satellite System
GOS	Geospatial One-Stop
GPS	Global Positioning System
HAZMAT	Hazardous Materials
HIFLD	Homeland Infrastructure Foundation-Level Database
HLS	Homeland Security
HSEEP	Homeland Security Exercise and Evaluation Program
HSGP	Homeland Security Grant Program
HSIN	Homeland Security Information Network
HSIP	Homeland Security Infrastructure Protection
HSPD	Homeland Security Presidential Directive
HSPF	Hydrologic Simulation Program FORTRAN
HVRI	Hazards and Vulnerability Research Institute
IA	Implementing Authority
IaaS	Infrastructure as a Service
IAIP	Information Analysis and Infrastructure Protection
iCAV	Integrated Common Analytical Viewer
ICS	Incident Command System
ID	Identification Number
IICD	Infrastructure Information Collection Division
IMSL	International Mathematics Subroutine Library
IMT	Incident Management Team

IPM	Integrated Pipeline Management
IT	Information Technology
LEO	Law Enforcement Officer
LNG	Liquefied Natural Gas
MDT	Mobile Data Terminal
MEMA	Mississippi Emergency Management Agency
MGRS	Military Grid Reference System
MOSS	Map Overlay and Statistical System
MOU	Memorandum of Understanding
MPC	Mobile Positioning Center
MSA	Metropolitan Statistical Area
MSAG	Master Street Address Guide
NAD-83	North American Datum 1983
NCAS	Non-Call-Path Associated Signaling
NENA	National Emergency Number Association
NextGen 911	Next Generation 911
NFA	National Fire Academy
NFPA	National Fire Protection Association
NGA	National Geospatial-Intelligence Agency
NGAC	National Geospatial Advisory Committee
NGDA	National Geospatial Data Asset
NGO	Nongovernmental Organization
NIEM	National Information Exchange Model
NIMS	National Incident Management System
NIPP	National Infrastructure Protection Plan
NIST	National Institute of Standards and Technology
NOAA	National Oceanic and Atmospheric Administration
NPG	*National Preparedness Guidelines*
NPMS	National Pipeline Mapping System
NRP	National Response Plan
NSDI	National Spatial Data Infrastructure
NSGIC	National States Geographic Information Council
NSSE	National Security Special Event
NTSB	National Transportation Safety Board
NTT	Nippon Telegraph and Telephone
NWCG	National Wildfire Coordinating Group
OMB	Office of Management and Budget
OPSG	Operation Stonegarden
OVAP	Occupancy Vulnerability Assessment Profile
PaaS	Platform as a Service
PDD	Presidential Decision Directive
PDE	Position-Determining Entity

PHMSA   Pipeline and Hazardous Materials Safety Administration
PII   Publicly Identifiable Information
PL1   Programming Language-1
PLR   Plant Location Record
POC   Point of Contact
POP   Point of Presence
PSAP   Public Service Access Point
RFI   Request for Information
RFID   Radio Frequency Identification
RFP   Request for Proposal
RFQ   Request for Quotation
RNC   Republican National Convention
ROADIC   Road Administration Information Center
ROADIS   Road Administration Information System
ROI   Return on Investment
ROW   Right-of-Way
SaaS   Software as a Service
SCADA   Supervisory Control and Data Acquisition
SDSFIE   Spatial Data Standard for Facilities, Infrastructure and
               Environment
SEOC   State Emergency Operations Center
SERT   State Emergency Response Team
SFI   Strategic Foresight Initiative on the Critical Infrastructure
SHSP   State Homeland Security Program
SLAMM   Sea Level Affecting Marshes Model
SMAC   State Mutual Aid Compact
SOC   Standard of Cover
SOP   Standard Operating Procedure
SoVI   Social Vulnerability Index
SRP   Salt River Project
SYMAP   Synagraphic Mapping System
TCL   Target Capabilities List
THIRA   Threat and Hazard Identification and Risk Assessment
TIGER   Tampa Information and Geographical Event Resources
TIGER   Topologically Integrated Geographic Encoding and Referencing
TPD   Tampa Police Department
TSA   Transportation Security Administration
TSP   Telecommunications Service Priority
TSSDS   Tri-Services Spatial Data Standard
TUMSY   Total Utility Mapping System
UASI   Urban Areas Security Initiative
UN   United Nations

USFA	U.S. Fire Administration
USFWS	U.S. Fish and Wildlife Service
USGS	U.S. Geological Survey
USNG	U.S. National Grid
UTM	Universal Transverse Mercator
VoIP	Voice-over-Internet Protocol
WGS-84	World Geodetic System, 1984 revision
WPS	Wireless Priority Service

1

Introduction

1.1 Disasters

Natural processes constantly bring changes to the earth. Floods and tsunamis, earthquakes and volcanoes, hurricanes and tornadoes, meteors and solar flares, and other natural events affect the planet daily. When those events take place and cause substantial loss of life or damage to property, we call them disasters.

An earthquake with a magnitude of 8.6 occurred in 2012 in the Indian Ocean, about 435 km southwest of Banda Acheh, Indonesia. No deaths were reported in conjunction with this event, which caused a small, 17-cm high tsunami. This distinguishes it from the 2004 earthquake (magnitude 9.1) in the same area that resulted in more than 230,000 deaths, many caused by waves up to 30 m in height. This later event would be considered a natural disaster, while the former was a natural event.

The "one-two punch" of earthquakes and tsunamis can have a profound near-term impact on human life, but also, through the damage caused to dwellings and other property, an impact on the quality of life for many years afterward. The 2011 earthquake (magnitude 9.0) in the Tohoku region of Japan, and the associated tsunami waves that reached 40 m in height, caused the death of more than 15,000 people. However, these events also caused the complete collapse of 128,808 buildings, the partial collapse of 269,871 other buildings, and significant damage to an additional 740,185 buildings, the majority of which (more than 680,000) were homes (National Police Agency of Japan, 2013).

This Great East Japan Earthquake left more than 4 million people without electricity and more than 1.5 million people without potable water. Roads, bridges, dikes, and railways were damaged, hindering repair efforts. The Fukushima Daiichi Nuclear Power Plant complex sustained well-documented damage, forcing the evacuation of residents living in the area. The insurance industry estimated the insurance losses associated with the earthquake alone to approach $35 billion (Hennessy-Fiske, 2011), while the World Bank estimated the total economic damages to be as great as $235 billion (Kim, 2011).

TABLE 1.1

CRED Summary of Natural Disasters

	1974–1978	1979–1983	1984–1988	1989–1993	1994–1998	1999–2003
Africa	88	113	128	107	149	333
Americas	99	199	255	319	320	475
Asia	220	336	353	482	449	726
Europe	43	108	136	144	134	288
Oceania	47	56	57	64	64	75
World	497	812	929	1,116	1,116	1,897

Source: Guha-Sapir, D., et al., *Thirty Years of Natural Disasters 1974–2003: The Numbers,* Centre for Research on the Epidemiology of Disasters, Presses Universitaires de Luvain, Brussels, 2004, p. 80.

There is no single measure of disasters. The Centre for Research on the Epidemiology of Disasters (CRED) maintains the Emergency Events Database (EM-DAT), a worldwide database on disasters, on behalf of the United Nations. For an event to be deemed a disaster and included in the EM-DAT database, it must have caused at least one of these consequences: (1) the reported death of 10 or more people, (2) adverse effects on 100 or more people, (3) a declaration of a state of emergency, or (4) a call for international assistance (Hoyois et al., 2007, p. 15).

Using the CRED definition, Table 1.1 summarizes natural disasters worldwide for a recent 30-year period.

Recognizing the importance of data sharing, the European Space Agency and the Centre National d'Etudes Spatiales (CNES) created an international charter in 2000. The charter's goal is to provide "a unified system of space data acquisition and delivery to those affected by natural or man-made disasters" (International Charter, 2013a). The charter's resources are activated upon the request of an authorized user. Although this is a more restrictive definition, it does provide a reasonable indicator of the perceived impact of an event from the perspective of the affected nation. Table 1.2 summarizes a recent 10-year period of international charter data.

Just as natural events or processes can become disasters, so also can the hazards associated with human actions. In contrast with natural disasters, man-made disasters are a consequence of human action (or inaction). This category includes technology-oriented events, such as the collapse of structures (bridges and buildings, for example), industrial accidents, and nuclear power plant accidents. It also includes acts of war and terrorism.

The category of man-made disasters includes train crashes, such as the 1998 Enschede, Netherlands, train crash that killed 101 people and injured 105. In another tragic example, the 1977 Granville Railway Bridge collapse in Sydney, Australia, 83 people died and 210 were injured. Both of these disasters involved railways and bridges, demonstrating the interconnected nature of many man-made disasters.

TABLE 1.2

International Charter Activations

Year	Total	Natural	Man-Made	% Natural
2003	17	17	0	100.0
2004	21	20	1	95.2
2005	25	25	0	100.0
2006	24	21	3	87.5
2007	43	40	3	93.0
2008	39	39	0	100.0
2009	40	40	0	100.0
2010	51	50	1	98.0
2011	32	31	1	96.9
2012	40	40	0	100.0
Total	332	323	9	97.3

Source: International Charter, Charter Activations, 2013, http://www.disasterscharter.org/web/charter/activations.

The poison gas leaks on December 3, 1984, from the Union Carbide pesticide factory in Bhopal, India, certainly constitute a man-made disaster. Called by some the worst industrial accident in history, the disaster took the lives of 3,800 people according to Union Carbide, although local authorities cleared 15,000 bodies immediately afterward (Long, 2008). As many as 50,000 people were injured, many of them permanently.

In some cases, the distinction between types of disasters is blurred and, at least to the victims, irrelevant. For example, in 2008 China experienced a magnitude 7.9 earthquake in Sichuan Province. China's building codes for schools were described by competent authorities as adequate and well defined. Nevertheless, more than 7,000 schoolrooms collapsed, reportedly due to poor construction practices, causing the death of thousands of children (BBC News, 2008). This would be considered an example of a natural disaster, the effects of which were amplified by human actions and which therefore also became a man-made disaster.

There is perhaps no clearer example of terrorism as a man-made disaster than the September 11, 2001, attacks by the Islamic terrorist group al-Qaeda. Conducted using skyjacked airplanes, the attacks on the Twin Towers in New York City, the Pentagon in Arlington, Virginia, and a third, unconfirmed target caused the immediate death of 2,977 people and the 19 hijackers. The long-term health effects from the collapsing towers and fire, including respiratory system damage, are still unknown.

Whether natural or man-made, or through a combination of events, disasters affect lives by causing death and injury to people and by causing damage to the infrastructure upon which people depend to sustain life. These impacts can be prevented in some cases, and can be minimized or mitigated in many other cases. This text considers the role of geographic information systems (GIS) technology as an agent for mitigation.

1.2 Definition of Critical Infrastructure

Before we can define *critical infrastructure*, we must define what we mean by the term *infrastructure*. A typical dictionary definition identifies *infrastructure* as "the basic physical and organizational structures and facilities (for example, buildings, roads, power supplies) needed for the operation of a society or enterprise: the social and economic infrastructure of a country" (*Oxford English Dictionary*, online). Traditionally, the term referred to physical structures that were needed and used in everyday life.

Every 4 years, the American Society of Civil Engineers (ASCE) produces a report card that depicts the condition and performance of the nation's public infrastructure. The ASCE definition of infrastructure includes several categories, such as water and environment (dams, drinking water, hazardous waste, levees, solid waste, and wastewater), transportation (aviation, bridges, inland waterways, ports, rail, roads, and transit), public facilities (public parks and recreation and schools), and energy.

The ASCE emphasis on public infrastructure is significant. For example, many ports and transit systems are managed by government, but many others are managed and partially owned by the private sector. Although rail stations are often public buildings, the track and supporting infrastructure associated with railways, as well as the rolling stock (that is, engines and cars of all types), are owned by private entities. The infrastructure used to support the energy category is almost exclusively held by private sector interests or as user cooperatives. Indeed, some sources consider 85% of the U.S. infrastructure to be held by the private sector.

The distinction between public and private investment is important to understanding the levels of investment in, and methods of using, infrastructure. This, in turn, is important when planning to construct and maintain specific components of the nation's infrastructure. However, it becomes a moot point when we consider the extent to which users ignore ownership and focus on the key question, is the infrastructure dependably available? Put simply, infrastructure users don't care who owns things; they only care that the infrastructure is available and working.

In recent years, the definition of infrastructure has expanded to include social structures and, in some cases, cyber structures, while maintaining the emphasis on everyday use. The number of definitions of infrastructure is limited only by the number of organizations, agencies, and government departments that have addressed the subject. Several authors have noted the inconsistencies and incompatibilities that characterize many of these definitions.

Some of the definitions are so expansive that they become less useful than other, more specific definitions. In response, Fulmer (2009, p. 32) offers this concise definition of infrastructure: "the physical components of interrelated systems providing commodities and services essential to enable, sustain, or enhance societal living conditions." The interrelated nature of these systems is particularly significant when we consider the problem of infrastructure protection, a point to which we will return throughout this text.

The term *critical* also has several meanings, but we focus on one in this book. For our purposes, something is judged critical if it is extremely important to the success (or failure) of some present or future human activity. Critical infrastructure, in simplest terms, is infrastructure (public or private) whose availability is *extremely important* to the successful operation of a civilized society.

This definition was elaborated by President William Clinton in Presidential Decision Directive 63 (Clinton, 1998, pp. 1–2):

> Critical infrastructures are those physical and cyber-based systems essential to the minimum operations of the economy and government. They include, but are not limited to, telecommunications, energy, banking and finance, transportation, water systems and emergency services, both governmental and private. Many of the nation's critical infrastructures have historically been physically and logically separate systems that had little interdependence. As a result of advances in information technology and the necessity of improved efficiency, however, these infrastructures have become increasingly automated and interlinked. These same advances have created new vulnerabilities to equipment failure, human error, weather and other natural causes, and physical and cyber attacks. Addressing these vulnerabilities will necessarily require flexible, evolutionary approaches that span both the public and private sectors, and protect both domestic and international security.

Given the breadth of the official definition of *critical infrastructure*, it is hardly surprising that separate systems existed to document and maintain pieces of infrastructure. Individual telecommunications companies, electric utilities, gas pipelines, railroad companies, and others have managed their corporate assets as capital investments made on behalf of their shareholders or cooperative members. In some cases, competitive issues would have impeded the exchange or sharing of information between these commercial entities.

Similarly, municipalities, counties, regional agencies, and state governments have collected and maintained information about investments made on behalf of the public, including investments in water, wastewater, storm sewers, road, and other facilities. It is worth noting that within this hierarchy of governments, there are significant differences in legal responsibilities

that may affect data sharing. For example, in many states the county governments are responsible for managing public service access points (PSAPs),
also known as 911 call centers. Adjacent counties would not necessarily perceive a need to share data during normal operations, a perception that might
have unfortunate consequences if an incident occurs near a shared border
and there is a miscommunication about responsibility for responding.

Similarly, state governments fund, and collect information about, state
roads and highways. However, municipalities manage the construction and
maintenance of local streets, typically at a much finer level of granularity
due to the related need to manage access to rights-of-way, underground utilities, and public works such as storm drains, as well as street addresses for
the dispatch of police, fire, and paramedic services. These different scales of
record keeping affect the ability to share data by creating incompatibilities
in core computer databases.

Despite these differences, and as noted by President Clinton, information
technology has advanced to the point that data sharing or, with increasing frequency, access to data via web or data services is commonplace. This
creates opportunities for interagency analysis and data use. However, the
interconnectedness of systems can also create vulnerabilities. In either case,
critical infrastructure must be protected.

1.2.1 Critical Infrastructure Sectors

As a result of Presidential Policy Directive 21 and Homeland Security
Presidential Directive 7 (http://www.dhs.gov/critical-infrastructure-sectors),
16 critical infrastructure sectors were defined:

1. The chemical sector supports a broad set of critical infrastructure
   sectors that includes consumer materials, the pharmaceutical industry, and basic, agricultural, and specialty chemicals.

2. The commercial facilities sector comprises commercial and residential real estate, gaming facilities, lodging, media and entertainment
   services, outside events and facilities, public meeting facilities, retail
   centers, and sports facilities.

3. The private sector owns and operates a majority of communications infrastructure, including the financial services, emergency
   response providers, energy suppliers, and information technology
   services providers.

4. The manufacturing sector includes a number of vital industries,
   such as electric equipment, machinery, metal, and transportation
   equipment manufacturing.

5. The dam sector includes an inventory of over 87,000 dams across
   the United States, of which roughly 65% are privately owned. Dams

support a variety of sectors, including emergency services, energy, food and agriculture, transportation systems, and water and wastewater systems.

6. The defense industrial base sector involves development, production, delivery, and maintenance of military weapons systems, subsystems, and components for each of the branches of the U.S. military.

7. The emergency services sector involves the prevention, preparedness, response, and recovery during intentional and unintentional man-made events and natural disasters. This sector plays a crucial role in protecting the critical infrastructure sectors by providing emergency services, management and medical services, fire response, law enforcement, and a variety of public works functions.

8. The energy sector includes three major segments for the generation and delivery of electricity, the production and transmission of petroleum, and natural gas. The electricity segment contains over 6,400 power plants powered by coal, gas, oil, water, or nuclear sources of fuel.

9. The financial services sector represents a vital segment of critical infrastructure that provides a wide range of banking-related products and services needed by both government and private entities during and after a major disaster with prolonged outages.

10. The food and agriculture sector is largely made up of privately held farms, restaurants, and food manufacturing, processing, and storage facilities. There are a variety of dependencies with a number of other sectors, including energy, financial services, transportation systems, and water and wastewater systems.

11. The government facilities sector includes owned and leased buildings and facilities located in both the United States and internationally. This sector also includes cyber elements that contribute to the protection of government and military assets. Other subsectors include public and private educational facilities and national monuments.

12. The healthcare and public health sector is largely privately held and safeguards the nation's economy from natural disasters, terrorist attacks, and major outbreaks from infectious disease. This sector is dependent on other sectors for the continuity of services and operations involving communications, emergency services, energy, food and agriculture, information technology, transportation systems, and water and wastewater systems.

13. The information technology sector is core to the nation's economy, physical and cyber security, and public health and safety. This sector is owned and operated by a combination of public and private entities. Information technology is highly interdependent and

interconnected with other critical assets that are vital to coordinating and performing emergency management preparedness and protection functions.

14. The nuclear reactors, materials, and waste sector provides approximately 20% of the nation's electrical generation by approximately 100 commercial reactors. This sector includes nuclear power plants; nonpower nuclear reactors used for research, testing, and training; nuclear equipment manufacturers; radioactive materials and nuclear fuel facilities; decommissioned reactors; and transportation, storage, and disposal of nuclear and radioactive waste. The sector is dependent on the chemical, energy, healthcare and public health, and transportation system sectors.

15. The transportation systems sector is responsible for the movement of people and goods across the nation and overseas. The transportation systems sector consists of aviation, highways, maritime, mass transit, passenger and freight rail, pipelines, and shipping.

16. The water and wastewater systems sector includes approximately 160,000 public drinking water facilities and systems, and over 16,000 publicly owned wastewater treatment systems across the United States. Vital services such as firefighting and healthcare and other interdependent sectors, including energy, food and agriculture, and transportation systems, would be severely impacted if this sector is disrupted for any extended period of time.

1.3 Critical Infrastructure Protection

1.3.1 Geographic Nature of Crisis and Emergency Response

This text provides fundamental concepts and experiences related to the emergency response and disaster management communities because they are considered critical infrastructure and because they are requisite elements for the protection of all other types of critical infrastructure. A critical lesson learned from Hurricane Katrina was that there are significant disconnections between the emergency response and management community, geospatial professionals, and infrastructure owners and stakeholders. The most basic misunderstandings arise from a lack of awareness about the roles and capabilities of each group as they pertain to critical infrastructure protection (CIP).

From an emergency responder perspective, all disasters and crises are local. They begin at a point, radiate outward geographically, and grow in complexity as the consequences of an incident intensify with time. There is a direct relationship between the onset of an incident, its final geographic scope, and the degree to which life and property are preserved. However, the fundamental tenets of geography or the potential practical use of GIS is not foremost in an emergency responder's mind. Rather, the strategies and tactics pertaining to resolving the crisis are at the fore, and making a map seems as easy as drawing and labeling some lines on a page.

Conversely, the application of geographic concepts and GIS to crisis or infrastructure protection is often obvious to a geospatial professional. However, extinguishing a fire may seem as simple as putting the "wet stuff on the red stuff." One of the key challenges this work seeks to remedy is bridging this gap by creating a mutual understanding of both crisis response and management and geography.

Geography also plays a significant role in determining risk: the types of crises faced, how events progress, and the resources available to address the associated problems. For example, inland cities are not endangered by tsunamis, and the equipment and resources used to mitigate tsunami damage are rarely located far from open seas and oceans. Similarly, it is highly unlikely that an aircraft crash/rescue firefighting truck would be found in a fire station located far away from an airport.

Advanced emergency medical services are more likely available in major urban centers than in remote rural areas (although cyber or telemedicine may mitigate this deficiency in the future). Wildfires burn more readily uphill than downhill, whereas floods move downhill rather than uphill, affecting large-incident strategy. Geography may also influence tactics at relatively small events, such as home fires, where single-family dwellings built with fire-resistant materials and sprinkler systems are more frequently found in newer communities, and structures that are more flammable are typically found in older municipalities.

Expressed simply, a fire will almost always start with a spark, arc, or other point source of ignition. The fire will spread geographically, governed by the properties of thermodynamics, with the availability of fuel, oxygen, and heat. Its extent will expand until one or more of those three elements is exhausted or removed through the mitigating actions of a first responder.

As the size of the fire grows, the underlying complexity of the event increases until a point is reached at which the immediate resources of the first responders are exhausted, the event has become uncontrollable, and additional aid is required. This is the start of a disaster if left unchecked and the pattern, whether an earthquake radiating outward from its epicenter or the surge of floodwaters from a breached dam, is constant.

1.3.2 Basic Concepts in Emergency Response

The ordered priorities of emergency responders should be clearly understood: initial actions must be to mitigate loss of life, contain the incident, and preserve property. The actions taken by first responders to accomplish each of these priorities are based upon a balanced evaluation of risk and reward conducted by the incident commander. His or her evaluation is rooted in situational awareness, training, knowledge of the capabilities of the resources at hand, and often decades of experience. A familiar adage in the fire service is "risk a lot to save a lot, risk a little to save little."

For illustrative purposes, imagine a gas leak in a distribution main pipeline near a typical two-story, eight-unit apartment building. Responders will focus immediately on evacuation of the area, thereby reducing the immediate risk to life. Efforts to harness the leak will occur as and when additional resources arrive.

If there were a subsequent explosion and fire that fully engulfed one unit in the apartment building and began to spread, resources would be assigned to search adjacent units and rescue obvious survivors before the deployment of one or more hose lines to contain the fire. Under ideal circumstances, enough resources would be sent to the scene of the incident at the time of dispatch, when the incident is just a leak, in anticipation of the potential for explosion and fire and such that all three actions (evacuation, search and rescue, and firefighting) may be initiated simultaneously and with success.

Note that the adjacent units represent areas with the highest probability of successfully finding a potential victim alive—*not* the apartment fully engulfed in flames. As additional resources arrive, firefighters will make entry into the apartment of origin to begin extinguishment and conduct a search of the fire apartment. This allocates risk to firefighter lives to the locations most apt to result in a successful rescue.

In the same vein, if the location of a potentially trapped person within the fire apartment is known or a survivor were to appear miraculously at a window filled with fire, tremendous risk would be undertaken to affect the rescue, as the potential for reward increases dramatically. In fact, the fire might be allowed to spread while an attempt at rescue is undertaken as life ranks higher in the overall priority scheme.

The management of priorities by emergency responders may at times seem counterintuitive. The example of a fire on board a large ship with many passengers serves to illustrate this conundrum. If there was a fire in the boiler room of a ship and one of four firefighters attacking the raging blaze was to fall and break a leg, the priority would be to continue the firefight and potentially sacrifice the downed firefighter. The rationale behind such a decision would be that were the fire not controlled quickly, the ensuing explosion would kill all and not just one.

The point herein is not to generalize how responders may react to a given scenario briefly described in a book about critical infrastructure protection, but rather describe some of the underlying thought processes used by first responders to control an incident involving infrastructure. Such a basic understanding should play a role in the geospatial products and services made available to responders.

For example, consider the possible effects this leak and any consequent explosions or fires might have on nearby critical infrastructure. If the apartment building was located adjacent to a hospital or power plant, responders would reevaluate the circumstances and adapt their behavior to accommodate this situation.

It is important to understand that an initial emergency response is often dictated by the worst possible scenario, yet limited by sensible risk management. Again illustrating by example, a fire department receives a call that an automatic fire alarm has activated in a water treatment plan with large stores of organic peroxides. The overwhelming majority of such alarms are due to malfunctioning detectors, and occasionally they will indicate the presence of smoke or a small fire.

In the rare instance, a disaster may be brewing. Yet, the initial response may be limited to two engines, a ladder company, and a chief officer. This strikes a careful balance that measures the risk associated with the incident at hand with respect to the risk of leaving the entire community without available responders should the entire department be sent on the initial alarm.

This is a reasonable and prudent approach, but one that can go quite wrong, as was the case on November 18, 2014, at a water treatment plant in Santa Paula, California. In this instance, a truck delivering organic peroxides caught fire on site and several firefighters were injured, evacuations of nearby neighborhoods conducted, and a major highway shut down. The fire was allowed to burn itself out, because the spread of contaminants through the application of water would have posed a greater risk.

Were an unknown liquid chemical leak discovered by arriving units, another tenet of emergency response would be employed—scalability. Emergency response organizations are designed in a largely modular fashion. As an incident grows in size and complexity, the quantity of specific resource types is increased.

In the water treatment plant example cited above, additional engines, hazardous materials (HAZMAT) teams and equipment, law enforcement officers, emergency medical services (EMS), and leadership elements would be added to the incident, as deemed appropriate by the incident response commander. As the incident is brought under control, the number of units on scene is scaled back while maintaining a reasonable margin of safety. This understanding of risk becomes more formalized when applied to emergency management where scenarios play out at much larger scales.

1.3.3 Risk Assessment Methodology

The Federal Emergency Management Agency (FEMA) defines risk as *Asset value × Hazard or threat rating × Vulnerability*. Each term may be ranked numerically; typically, each might be assigned a value between 0 and 10. The terms are then combined to determine risk.

For example, a child day care center may have an asset value of 9 because children are our future. The same center has a threat rating of 3 because the facility is exposed to few threats or hazards. Finally, the center has a vulnerability rating of 2 because the building is built partially underground, of steel-reinforced concrete, and supplied with enough peanut butter, jelly, and bread to last 6 weeks. The resultant numerical risk would then be calculated as $9 \times 3 \times 2$, or 54.

When compared across multiple aspects of a community, a committee guiding the risk assessment process may decide that risk factor scores below 60 are considered low, scores from 61 to 175 are considered moderate, and those with scores greater than 175 are high. While this method was originally developed by FEMA, and is more typically used for terror mitigation, the approach is sound and its principles applied in varying fashions throughout the United States.

An asset is defined as a resource of value requiring protection. An asset can be tangible, for example, buildings, facilities, equipment, activities, operations, and information, or intangible, for example, manufacturing processes, customer information, and a company's reputation. The value of an asset is measured in terms of the negative impact that would be caused by the incapacity or destruction of that asset. It is important to note that for the purposes of mitigation planning for natural hazards, and with the exception of critical infrastructure and key resources, we identify assets as groups or entities rather than individual structures or people.

In practical terms, some entities might appear in multiple categories. For example, a school may be considered a people-based asset, a piece of critical infrastructure, and a secondary response asset when used as an emergency shelter. For these cases, a higher asset value is assigned due to these multiple roles and to the importance of the facility to the community.

Threats and hazards, for the purpose of an initial planning period, are limited to natural events or failure of significant man-made infrastructure. Each hazard is again assigned a range of numeric values related to the perceived probability and severity of a hazard or threat. The overarching idea is to produce a hazard ranking that may be incorporated in the formula for risk ($Risk = Asset\ value \times Threat \times Vulnerability$).

Vulnerability is defined by FEMA as any weakness that can be exploited by an aggressor or, in a nonterrorist threat environment, could render an asset susceptible to hazard damage. Within a community, assets share a common geography, and thereby the probability of any one or combination of threats affecting any one or combination of assets is equally likely, with few exceptions (flood, for example, is related to elevation, which may vary). The degree to which any asset may be affected by any threat is dependent upon the degree to which the threat is manifested—low, medium, or high intensity.

1.3.4 Fundamentals of Emergency Management

Several variables are examined to determine if emergency management organizations are likely to become involved at the request of the incident commander. The first variable considered is geographic magnitude: Does the event cover a large geographic area, requiring a multijurisdictional response? The second variable is technical complexity: Is the event exceedingly complex, involving multiple disciplines (for example, police, fire, emergency medicine, environmental specialists, and utility companies)?

The third variable is the complexity of logistical planning. The fourth variable is the duration: Is the event likely to occur over long periods of time (days or weeks)? The primary role of an emergency manager and the affiliated organization is to provide support and coordination for large incidents. As with emergency response, emergency management is also scalable.

In the example of the water treatment plant, a prudent incident commander would request emergency management support. Such support might entail providing a rest and recuperation area for firefighters and HAZMAT teams and arranging for shelter for displaced residents. It may also include the provision of a public information officer to coordinate media affairs.

The support might also extend to arranging for environmental quality specialists to sample the air downwind from the fire and determine if additional evacuations are needed or similar activities. Like emergency response, emergency management is a scalable activity. It is modular in nature, drawing upon specific resources that are typed (classified) by their capacity and capabilities.

To be clear, emergency management organizations are not first responders. They are not typically trained or equipped to respond to the immediate scene of an event, and they are not ultimately in charge of how an event is resolved. Whereas emergency responders are typically involved with an event until the cessation of the immediate threat and potential for loss of life and property, emergency management organizations must contend with a complete life cycle of planning and preparedness, mitigation, response, and recovery on a far larger scale.

Emergency management organizations, like the military and emergency responders, exist within the duality of peacetime and activation for an event. Military and emergency responders tend to focus on skill development and scenario training during peaceful times, and so do emergency management organizations, but with some interesting twists.

By nature, emergency management must cope with a broad spectrum of potential incidents, such as large public events similar to the Republican National Convention described later in this work in Chapter 6, natural disasters, terrorist attacks, infrastructure failures (such as large-scale blackouts), and nearly anything else that a Hollywood producer could imagine for a gut-wrenching cinematic thriller. Thus, emergency managers tend to focus more on planning and mitigation activities, as those are the areas that present the greatest opportunity for return on investment or diminishing risk.

The planning and preparedness phases of emergency management involve two major activities. The first activity comprises identifying assets requiring protection from a wide variety of threats and hazards, gaining an under-standing of the psychological and absolute costs were they to be lost during a crisis, assessing what actions may be prudently taken to mitigate their dam-age and loss, and creating a plan that sets those mitigating actions into play.

The second activity involves the creation of a generalized "playbook" by which emergency response resources are catalogued, including the methods by which they may be contacted, activated, employed, and paid when they are needed. The playbook also includes a detailed description of who will fulfill what leadership roles and responsibilities should a crisis occur. The first set of activities should result in the creation of a hazard mitigation plan, whereas the second are used to create a comprehensive emergency manage-ment plan.

More often than not, the bulk of personnel working in emergency man-agement are engaged in some type of planning or training activity during noncrisis times, though oversight of mitigation activities is not uncommon. Hazard mitigation encompasses a large variety of activities that are typically based upon reducing risks associated with threats to specific assets. This is a geographic problem susceptible to analysis through GIS.

For example, severe thunderstorms are more probable in the midwestern United States than they are in the desert areas of north-central Nevada. Thus, tree trimming as a means of preventing power outages due to downed limbs is likely a higher-priority mitigation action in the Midwest than it would be in the desert. Likewise, the construction and maintenance of levee sys-tems along flood-prone rivers may be of a higher priority in the Sacramento, California, area than in other parts of the country.

Emergency management personnel may also be engaged in recovery from events that transpired in the recent past. Activities associated with recovery include providing support for citizens attempting to file claims for disas-ter relief funds and arranging for temporary long-term housing. They may also include arranging for, and supervising, contracts for debris removal and other hazardous conditions created by a disaster.

These emergency management personnel may also provide assistance to local government and citizens during the reconstruction period. In the United States, recovery activities typically fall within the domain of emer-gency management as long as a disaster declaration is in place at the local, state, or federal level.

The other duality of emergency management is that unlike emergency response, it is replicated at numerous levels of government. At the local level, emergency management is prone to providing coordinating efforts for responders and citizens. Local emergency management agencies will also liaise with the next higher level of governance, most often a state emergency management agency.

Likewise, state agencies will provide coordination in support of numerous local emergency management organizations while interacting with higher-level agencies such as FEMA. Thus, the emergency management life cycle is in continuous practice during both times of peace and times of crisis at all levels of governance.

Both emergency response and emergency management are highly regimented, predictably compartmentalized, and hierarchical in organization. Such a rigorous structure was borne from painful experience. Oakland, California, suffered from a rash of devastating wildfires in the mid-1980s. More than 100 lives were claimed, and more than 1,000 structures were lost.

After-action reports revealed that the losses were not as much from a lack of emergency responder resources and the intensity of the wildfires as from a lack of coordination, which would have allowed the right resources to be deployed at the right places and times, thereby reducing losses. As a result, the state of California created a flexible, modular system of incident management and required, by law, that all first responder and emergency management agencies be trained in its use. Communities failing to implement this management system faced a potential loss of state dollars for incident recovery funding.

Though conceptually adopted in numerous jurisdictions, a true national standard, now formally named the Incident Command System (ICS), did not come into effect until the passage of the Patriot Act in 2001. As in prior disasters, many emergency responders lost their lives when the World Trade Center collapsed on September 11, 2001, in part because of a lack of unified command and control (specifically, radio systems among the various responders were not interoperable).

The ICS, which is now used universally throughout the United States, follows the structure illustrated in Figure 1.1 (FEMA, 2013).

Typically, the first arriving emergency responder establishes and maintains command of an incident until he or she passes it to a more senior or otherwise designated responder. For small incidents, the incident commander may serve numerous roles. Conversely, for large incidents, roles may be assumed by teams. In all cases, the terminology used to describe positions and responsibilities is well defined and inflexible to minimize potential confusion, especially as it applies to the chain of command. While incident commanders may change throughout an incident, the position remains that of ultimate authority.

As shown, incident staffing is divided into two functional areas: command (for example, incident commander, public information officer, liaison officer, and safety officer) and general staff (for example, operations, planning, logistics, and finance/administration section chiefs). Command staff positions are responsible for the execution of goals and objectives established by the incident commander. General staff positions provide the resources

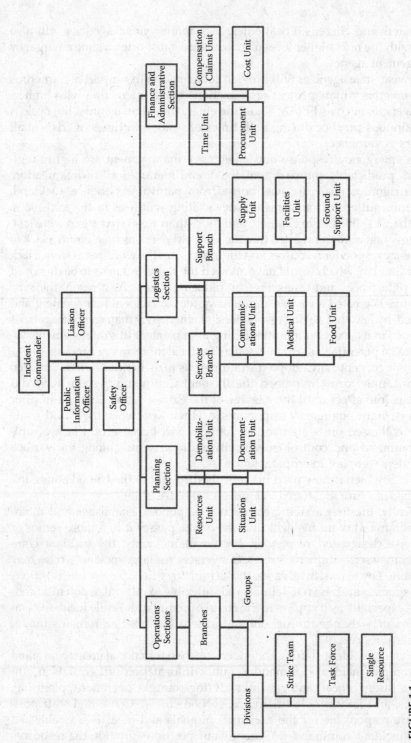

FIGURE 1.1
Incident command system.

required by command staff to accomplish their missions. Each group is of equal importance; the aphorism that "an army marches on its stomach" is no less true for emergency responders than for the military.

The drive for organizational consistency is maintained at the emergency management level as well. Here, support staffing is often delegated to emergency service functions (ESFs). Each ESF is described in Table 1.3, and their assigned areas of responsibility are shown in Table 1.4. While these tables describe federal-level organization, the general structure and assignments are replicated at lower levels of government.

Depending on the size of the organization, functions may be combined and managed by a few individuals or assigned to large, dedicated teams. Understanding that a common, consensus-driven set of standards governs emergency response and management is critical to successful integration of geospatial technologies for critical infrastructure protection.

TABLE 1.3

Emergency Service Function Descriptions

ESF	Scope
ESF 1: Transportation	Aviation/airspace management and control
	Transportation safety
	Restoration/recovery of transportation infrastructure
	Movement restrictions
	Damage and impact assessment
ESF 2: Communications	Coordination with telecommunications and information technology industries
	Restoration and repair of telecommunications infrastructure
	Protection, restoration, and sustainment of national cyber and information technology resources
	Oversight of communications with the federal incident management and response structures
ESF 3: Public works and engineering	Infrastructure protection and emergency repair
	Infrastructure restoration
	Engineering services and construction management
	Emergency contracting support for life-saving and life-sustaining services
ESF 4: Firefighting	Coordination of federal firefighting activities
	Support to wildland, rural, and urban firefighting operations
ESF 5: Emergency management	Coordination of incident management and response efforts
	Issuance of mission assignments
	Resource and human capital
	Incident action planning
	Financial management

(Continued)

TABLE 1.3 (Continued)

Emergency Service Function Descriptions

ESF	Scope
ESF 6: Mass care, emergency assistance, housing, and human services	Mass care Emergency assistance Disaster housing Human services
ESF 7: Logistics management and resource support	Comprehensive, national incident logistics planning, management, and sustainment capability Resource support (facility space, office equipment and supplies, contracting services, etc.)
ESF 8: Public health and medical services	Public health Medical Mental health services Mass fatality management
ESF 9: Search and rescue	Life-saving assistance Search and rescue operations
ESF 10: Oil and hazardous materials response	Oil and hazardous materials (chemical, biological, radiological, etc.) response Environmental short- and long-term cleanup
ESF 11: Agriculture and natural resources	Nutrition assistance Animal and plant disease and pest response Food safety and security Natural and cultural resources and historic properties protection and restoration Safety and well-being of household pets
ESF 12: Energy	Energy infrastructure assessment, repair, and restoration Energy industry utilities coordination Energy forecast
ESF 13: Public safety and security	Facility and resource security Security planning and technical resource assistance Public safety and security support Support to access, traffic, and crowd control
ESF 14: Long-term recovery	Superseded by the National Disaster Recovery Framework and no longer utilized
ESF 15: External affairs	Emergency public information and protective action guidance Media and community relations Congressional and international affairs Tribal and insular affairs

Source: Adapted from Federal Emergency Management Agency, ESF Annexes Introduction, 2008, to reflect roles as of 2014.

TABLE 1.4

Emergency Service Function Responsibilities

Agency	ESF-1: Transportation	ESF-2: Communications	ESF-3: Public Works and Engineering	ESF-4: Firefighting	ESF-5: Emergency Management	ESF-6: Mass Care, Emergency Assistance, Housing, and Human Services	ESF-7: Logistics Management and Resources Support	ESF-8: Public Health and Medical Services	ESF-9: Search and Rescue	ESF-10: Oil and Hazardous Materials Spill Response	ESF-11: Agriculture and Natural Resources	ESF-12: Energy	ESF-13: Public Safety and Security	ESF-15: External Affairs
Department of Agriculture	S	S	S		S	S	S	S	S	S	C/P/S	S	S	S
Forest Service	S	S	S	C/P	S	S	S	S	S	S	S	S	S	
Department of Corrections	S	S	S	S	S	S	S	S	S	S	S	S	S	S
Department of Defense	S	S	S	S		S	S	S	P	S	S	S	S	S
U.S. Army Corps of Engineers	S	S	C/P	S	S	S	S	S	P	S	S	S	S	S
Department of Education	S				S	S		S						S
Department of Energy	S		S	S	S	S	S	S	S	S	S	C/P	S	S
Health and Human Services	S		S	S	S	S	S	C/P	S	S	S	S	S	S
Department of Homeland Security	S	S	S	S	S	S	S	S	C/P	S	S	S	C	S
Federal Emergency Management Agency	S	P	P	S	C/P	C/P/S	C/P	S	C/P	S	S	S		P
National Communications System		C/P										S		
U.S. Coast Guard	S	S	S	S	S	S	S	S	P	P	S	S	S	
Housing and Urban Development	S					S								S
Department of the Interior	S	S	S	S	S	S	S	S	S	S	P/S	S	S	S

(Continued)

TABLE 1.4 (Continued)

Emergency Service Function Responsibilities

Agency	ESF-1: Transportation	ESF-2: Communications	ESF-3: Public Works and Engineering	ESF-4: Firefighting	ESF-5: Emergency Management	ESF-6: Mass Care, Emergency Assistance, Housing, and Human Services	ESF-7: Logistics Management and Resources Support	ESF-8: Public Health and Medical Services	ESF-9: Search and Rescue	ESF-10: Oil and Hazardous Materials Spill Response	ESF-11: Agriculture and Natural Resources	ESF-12: Energy	ESF-13: Public Safety and Security	ESF-15: External Affairs
Department of Justice	S				S	S		S	S	S	S		P/S	S
Department of Labor			S	S	S	S		S	S	S	S	S		S
Department of State	S		S		S			S		S	S	S		S
Department of Transportation	C/P		S		S	S	S	S		S	S	S		S
Treasury					S									S
Veteran's Administration			S		S	S	S	S					S	S
Environmental Protection Agency	S		S	S	S	S		S		C/P	S	S	S	S
Federal Communications Commission		S			S									S
General Services Administration	S	S	S		S	S	C/P	S	S	S	S			S
National Aeronautics and Space Administration					S								S	S
Nuclear Regulatory Commission					S							S		S
Office of Personnel Management					S		S							S
Small Business Administration					S	S								S

(Continued)

TABLE 1.4 (Continued)
Emergency Service Function Responsibilities

Agency	ESF-1: Transportation	ESF-2: Communications	ESF-3: Public Works and Engineering	ESF-4: Firefighting	ESF-5: Emergency Management	ESF-6: Mass Care, Emergency Assistance, Housing, and Human Services	ESF-7: Logistics Management and Resources Support	ESF-8: Public Health and Medical Services	ESF-9: Search and Rescue	ESF-10: Oil and Hazardous Materials Spill Response	ESF-11: Agriculture and Natural Resources	ESF-12: Energy	ESF-13: Public Safety and Security	ESF-15: External Affairs
Social Security Administration						S		S			S		S	S
Tennessee Valley Authority			S		S			S			S	S	S	S
U.S. Agency for International Development	S					S		S	S		S			S
U.S. Postal Service						S								S
Advisory Council on Historic Preservation														
American Red Cross					S	S		S			S			
Corporation for National Community Service														
Heritage Emergency National Task Force														
National Archives and Records Administration														
National Voluntary Organizations Active in Disaster						S								

Source: Adapted from Federal Emergency Management Agency, ESF Annexes Introduction, 2008, to reflect roles as of 2014.

Note: C = ESF coordinating agency. P = Primary agency. S = Support agency.

The importance of standards in both emergency management and emergency response cannot be overemphasized, as both fields are driven by standards. An electric operations manager at a power generation facility must follow strict protocols when bringing a generator at a power station back online after a regional power outage. A paramedic must follow strict rules of engagement when dealing with victims exposed to a lethal biological agent.

An emergency manager must meet threshold requirements before calling out certain resources. Failure to adhere to these standards comes with swift and severe consequences, ranging from often embarrassing and politically uncomfortable explanations about unnecessary expenditures to the grim reality of describing how a "freelance" approach at an incident resulted in a preventable death.

Adding weight to the significance of standards in this environment is the fact that they form the basis for all training and planning. Standards that at first glance may seem to be of little significance, such as how to don and wear personal protective equipment, build upon other standards to create a vetted and well-rehearsed approach to managing and bringing an incident under control. That this is a staid and difficult-to-change process is fundamental to its reliability and to the trust placed in the process by multiple regimented professions.

This statement should not be interpreted to mean that emergency responders and managers are not capable of swiftly employing creative solutions to complex problems. Rather, it means that the potential efficacy of a solution will be weighed against not only risk, but also the potential for success, using a well-known and documented procedure, before ranging "off script." When such a change does occur, it will still be executed within a larger framework of standards and tested thoroughly.

1.3.5 Geospatial Technologies and Emergency Response and Management

The integration of geospatial technologies with emergency management and emergency response confronts significant issues. Geographic information systems, remote sensing technologies, and the global navigation satellite system (GNSS), commonly called the global positioning system (GPS), are technical approaches to problem solving that have evolved and continue to evolve at a pace far faster than most approaches used in emergency response and emergency management.

As one seasoned fire chief once observed, "The fire service in the United States represents 331 years of tradition uninterrupted by progress." The time involved in vetting a technical solution deployed for field use, in comparison with the length of time a given version of the technology is supported by vendors, poses a significant challenge. For example, many E-911 mapping solutions only supported the use of Esri's (previously known as the Environmental Systems Research Institute) shapefile data format for

GIS data through 2008, even though the Esri geodatabase format was first employed in 2000.

While this challenge limits the use of geospatial technologies in high-risk environments that are close to the front lines, it does not invalidate the need for, and potential benefits of, geospatial technologies in other arenas of emergency response and management. The two most promising avenues for employing geospatial technologies in emergency response and management with respect to critical infrastructure protection are (1) situational awareness and information integration and (2) planning, modeling, and scenario building.

A variety of GIS-based technologies were used experimentally for situational awareness and information integration throughout the 1990s. However, they did not achieve widespread use within the U.S. federal government until the early 2000s with the rollout of the Integrated Common Analytical Viewer (iCAV). While technology previously existed to provide web-based information viewers, the challenges associated with providing varying levels of access to data feeds based on security clearance or subject matter interest were significant.

Esri was able to work with the U.S. federal government and overcome these challenges. This paved the way for future iterations of iCAV and the spread of similar approaches using other frameworks, most notably one established by the state of Alabama using the Google Earth platform called Virtual Alabama.

Perhaps the most significant accomplishment associated with the development of geospatial web viewers was that this approach allowed incident commanders and other leadership elements within the emergency management framework to view and interact with field reports and conditions using a map as the central means of presentation. The presentation of information in a geographic framework, as opposed to written reports and tables, represented a significant leap forward in their ability to rapidly understand and monitor an event as it unfolded.

In essence, these viewers brought geospatial awareness and insight to the emergency response and management community through a mechanism that was quick and easy for nontechnical users to grasp. As multiple similar approaches developed, the user community realized that the next set of critical challenges resided in the lack of consistently available data and an underlying ability to manage and share those data.

For example, an emergency manager could view the location of response assets slogging through neighborhoods destroyed by a hurricane overlaid upon aerial imagery. Now, however, that manager could also access population density data and topographic information and could query the status of underlying infrastructure elements, particularly at the neighborhood level. Thus, the need to view and interact with data came into sharp focus.

Coupled with other driving factors, the development of the Department of Homeland Security (DHS) Infrastructure Protection data set rapidly rose in significance and became a huge federally funded effort. The importance of obtaining up-to-date geospatial information about infrastructure and the challenges inherent in doing so also became clear.

More important than the need to view the location of critical infrastructure during a crisis is the need to integrate such information in the planning and modeling processes. Planning for and modeling a crisis prior to its occurrence is a high-value emergency management activity. This is because it allows for the creation of reasonable and prudent plans and the identification of mitigating actions that may be undertaken.

Since the adoption of the Hazus software tool by FEMA in 1997, geospatially driven modeling applications have become increasingly accessible to the emergency management community. In fact, the use of Hazus as the de facto model for simulating earthquake, flood, and cyclone risks and consequences is widely accepted. FEMA's description of Hazus clearly illustrates the importance of both the role of critical infrastructure data and modeling disasters on a geographic basis:

> Hazus is a nationally applicable standardized methodology that contains models for estimating potential losses from earthquakes, floods, and hurricanes. Hazus uses Geographic Information Systems (GIS) technology to estimate physical, economic, and social impacts of disasters. It graphically illustrates the limits of identified high-risk locations due to earthquake, hurricane, and floods. Users can then visualize the spatial relationships between populations and other more permanently fixed geographic assets or resources for the specific hazard being modeled, a crucial function in the pre-disaster planning process.
>
> Hazus is used for mitigation and recovery as well as preparedness and response. Government planners, GIS specialists, and emergency managers use Hazus to determine losses and the most beneficial mitigation approaches to take to minimize them. Hazus can be used in the assessment step in the mitigation planning process, which is the foundation for a community's long-term strategy to reduce disaster losses and break the cycle of disaster damage, reconstruction, and repeated damage. Being ready will aid in recovery after a natural disaster.
>
> As the number of Hazus users continues to increase, so do the types of uses. Increasingly, Hazus is being used by states and communities in support of risk assessments to perform economic loss scenarios for certain natural hazards and rapid needs assessments during hurricane response. Other communities are using Hazus to increase hazard awareness. Successful uses of Hazus are profiled under Mitigation and Recovery and Preparedness and Response. Emergency managers have also found these map templates helpful to support rapid impact assessment and disaster response. (Hazus website: http://www.fema.gov/hazus)

Hazus paved the way for other significant geospatially driven models to enter the emergency response and management community.

The integration of geospatial technologies within the Computer-Aided Management of Emergency Operations (CAMEO) is another significant example of a modeling tool. Begun in 1986, the CAMEO suite of programs was created to provide emergency responders with an easy-to-use computer database of hazardous materials and the ability to model basic risk factors associated with their involvement in a crisis. Its evolution included the incorporation of a geospatial element in the late 1990s, and it now allows users to perform plume modeling and subsequent geographic analysis using GIS software.

As with any model, the underlying quality of data that are used defines the quality of results. Emerging challenges for modeling suites such as Hazus and CAMEO are the increasing detail and resolution of available data and the prediction of interactions among hazards and infrastructure elements. Current versions can clearly demonstrate when infrastructure is at risk and in the path of crisis, but cannot predict the consequences of failure. More robust and timely data are required than those that are currently available.

The key message of this discussion is that crisis and emergency response are highly regimented and standardized activities that require vetted solutions. These professional communities are responsible for ensuring the continued safety and operability of infrastructure. They will not, and indeed cannot, conform to the methods and principles by which the geospatial profession as a whole operates.

Rather, the geospatial profession, particularly those areas that deal with critical infrastructure, must strive to build relationships with the emergency response community to ensure that the transfer of knowledge and information can occur and that better solutions can be built and tested. Wastewater and electric utility field operations staff should not meet the local fire chief for the first time when a major power outage has just occurred and there is no electricity to operate the pumping station that would provide water to extinguish a fire at a nearby high-rise apartment complex.

1.4 Infrastructure Interdependencies Model

1.4.1 Introduction to Interdependencies

The awareness that our nation's critical infrastructures are tightly interdependent in complex ways is more evident today than ever before. As shown by the 2005 failure of the levee system caused by Hurricane Katrina, the prolonged electricity blackout in the northeastern United States in 2003, and many recent infrastructure disruptions, failure of one type of infrastructure can directly and indirectly affect other types of infrastructures, affect large geographic regions, and send ripples throughout the national economy.

The recognition of infrastructure interdependences was highlighted in Presidential Decision Directive 63 (PDD 63) on critical infrastructure protection. Promoted by the 1995 bombing in Oklahoma City, PDD 63 was the culmination of an extensive study by the President's Commission on Critical Infrastructure Protection. That study exposed the emerging capability of exploiting electric, gas, transportation, water, wastewater, telecommunications, and banking and finance infrastructures.

The directive acknowledged that our national and economic security and viability depend on our critical infrastructure and the associated information technology that supports them. To ensure their protection and reliability, national committees were created for each infrastructure sector to research weaknesses and problems.

In their findings, the commission noted that the "mutual dependence and interconnectedness made possible by the information and communications infrastructure lead to the possibility that our infrastructures may be vulnerable in ways they never have been before" (President's Commission on Critical Infrastructure Protection, 1997). Not understanding how disruptions to one infrastructure could cascade to affect other components can worsen response and recovery efforts. It can also leave infrastructure owners and emergency response personnel unprepared to deal effectively with the impacts of such disruptions.

The commission also noted that understanding, analyzing, and sustaining the robustness and resilience of the critical infrastructure and their interdependencies requires appropriate modeling tools. These tools would be needed to assess the technical, economic, and security implications of technology and policy decisions designed to ensure their reliability and security.

Historically, interdependencies have been considered to be either physical or geographic in nature. An example of a physical interdependence is that the water supply infrastructure depends on electric power to operate its pumps, while at the same time the electric power infrastructure must have water to make steam and cool its equipment.

Geographic interdependencies arise when infrastructure components (for example, electric transmission lines, water pipelines, gas pipelines, and telecommunications cables) share common corridors, such as public rights-of-way and railway lines. This proximity increases the vulnerabilities to, and consequences from, disasters in the same geographic area.

The proliferation of information technology, increased use of automated monitoring and control systems, and growing reliance on the open marketplace for purchasing and selling infrastructure commodities and services have combined to link infrastructures in new and complex ways. As a result, they also have created new vulnerabilities. The dependence of the energy marketplace on the Internet and other e-commerce systems, and the complicated links to financial markets highlight the extent of cyber and logical interdependencies.

Rinaldi et al. (2001) classified infrastructure interdependencies as being one of four types: physical, cyber, geographic, or logical. Physical interdependencies involve disruptions that physically affect one or more other infrastructures. The risk of failure and deviation from normal operating conditions in one infrastructure may pose a function of risk in another infrastructure.

Cyber interdependencies occur when the operation of one infrastructure is dependent upon another infrastructure via information or communication connectivity. This is the type of complex system in which control of a networked system is dependent upon the transmission of information.

Geospatial interdependencies involve the physical proximity of one infrastructure to another. An event such as an explosion of a gas main in an urban area could create correlated disruptions with other infrastructures, such as water and electric services to a community.

Logical interdependencies mean that the state of one infrastructure is dependent upon another, due to some economic or political decision. An example of this is the logical interdependency between the availability of fuel and the number of vehicles using the transportation infrastructure.

As noted previously, infrastructures are geographically (that is, geospatially) interdependent if a local environmental event can create state changes in them. Geospatial interdependency occurs when elements of multiple infrastructures are in close proximity. Given this proximity, events such as an explosion or fire or a train derailment could create correlated disruptions or changes in these geographically interdependent infrastructures.

Such correlated changes are not due to physical or cyber connections between infrastructures; rather, they arise from the influence the event exerts on affected infrastructures at the same time. For example, an electrical line and fiber-optic cables hung under a bridge connect elements of the electric power, telecommunications, and transportation infrastructures. The interdependency in these cases is due to proximity; the state of one infrastructure does not influence the state of another.

Traffic across the bridge does not influence the flow of electricity or transmission of communications. Because of the close spatial proximity, however, physical damage to the bridge could create correlated disruptions in the electric power, communications, and transportation infrastructures. Some interdependencies and their effects on infrastructure operations are caused by natural events, whereas others result from human intervention and errors.

In the case of interdependencies, it is also important to note that the geospatial aspects of critical infrastructure are also scale, time, and scope dependent. Consider complexity of scope: when compared to our nation's size, Hurricane Ike affected a relatively small area in September 2008. The failure of local infrastructure (roads, bridges, and water systems) had the largest effect on response, yet occurred at the smallest geographic scale, and restoration will take the longest period of time.

Furthermore, failure of regional infrastructure (particularly electric utilities) during Hurricanes Rita and Ike proved to be the basis for a potentially overarching threat of national significance—the failure of up to 25% of the petroleum industry. In turn, this relates to a time issue: How long can we go without power for those refineries? This implies that the restoration period is likely to be much briefer than that for other local infrastructure. In other words, we must consider not only the first-order interdependencies of infrastructure, but also the corresponding second-order interdependencies of scale and time.

In comparison, Hurricane Sandy was truly astounding in its size and power. At its peak size, 20 hours before landfall, Sandy had tropical storm–force winds that covered an area nearly one-fifth the size of the contiguous United States. Sandy's area of ocean with wave heights of 3.65 m or greater covered an area of 3.6 million km^2—nearly one-half the size of the contiguous United States, or 1% of the earth's total ocean area. At landfall, Sandy's tropical storm–force winds spanned 1,518 km of the U.S. coast.

Sandy's huge size prompted high-wind warnings to be posted from Chicago to eastern Maine and from Michigan's Upper Peninsula to Florida's Lake Okeechobee, an area that was home to 120 million people. Sandy's winds simultaneously caused damage to buildings on the shores of Lake Michigan at Indiana Dunes National Lakeshore and toppled power lines in Nova Scotia, Canada, locations more than 1,900 km apart. The cascading effects from its scale in size and the loss of energy, water, and transportation infrastructure had staggering economic effects that were still affecting parts of the northeastern United States 1 year later.

Figure 1.2 illustrates examples of interdependent relationships among electric, water/wastewater, gas and oil, communications, and transportation infrastructures.

These complex relationships are characterized by multiple connections among infrastructures (Figure 1.3). The connections create a complex web that, depending on the characteristics of its linkages, can result in a cascading effect across multiple infrastructures that can impact communities like those affected by Hurricane Sandy.

One of the many lessons for major events is that it is impossible to adequately analyze or understand the behavior of a given infrastructure in isolation from the environment or other infrastructures. Rather, we must consider multiple interconnected infrastructures and their interdependencies in a holistic manner when planning for and responding to cascading failures.

1.4.2 Critical Infrastructure Cascading Failures

The interdependencies within an individual infrastructure network are often well understood. The focus of our thinking must be on the influence or impact that one infrastructure can have on another. The key effects to model and understand are the chains of influence that cross multiple infrastructure sectors and induce potentially unforeseen effects. As depicted in Figure 1.4, these chains, potentially composed of multiple interdependency types, constitute the connections between infrastructures. These particular connections represent the cascading consequence of a typical event.

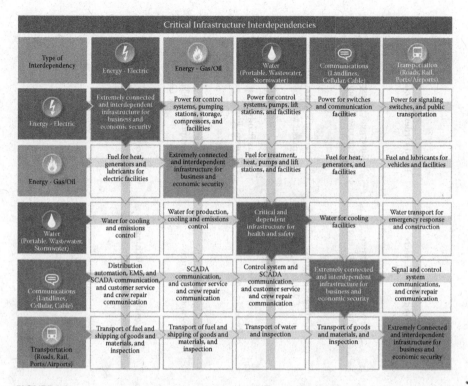

FIGURE 1.2
(See color insert.) Interdependency relationships. (*Source*: Geospatial Information and Technology Association, Birds of a Feather Committee's Report on Critical Infrastructure Interdependencies, 2008.)

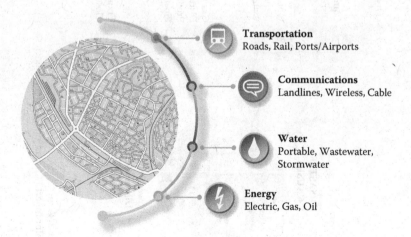

FIGURE 1.3
(See color insert.) Interdependency relationships.

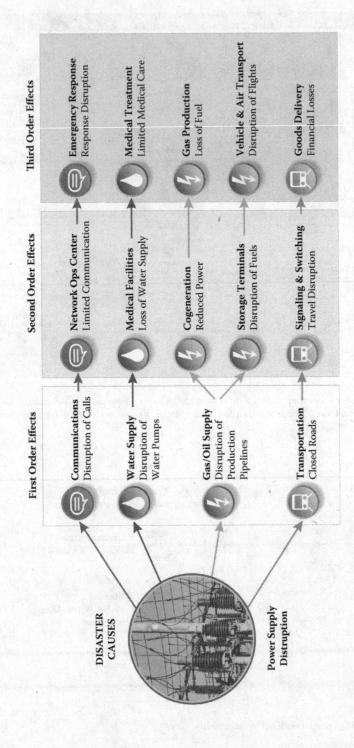

FIGURE 1.4
(See color insert.) Cascading interdependencies.

As discussed previously, a cascading failure occurs when a disruption in one infrastructure causes the failure of another infrastructure, which subsequently causes a disruption or escalating failure in the second infrastructure. For example, the disruption of a network within the water distribution infrastructure—the result of a construction accident—can result in a failure of an electric utility's cooling units located in the service area of the water system.

This event, in turn, can lead to a shortage of power generation in the area, which can cause power disruptions (a cascading failure from the water infrastructure to the electric power infrastructure). By extension, the electric power failure could lead to disruptions in other infrastructure, such as telecommunications and transportation.

In certain cases, a single event can cause two or more additional interrelated infrastructures to be disrupted at the same time. In other words, components with each network infrastructure fail because of some common cause or event. Components from multiple infrastructures could be affected simultaneously, either because the components share the same geographic location or because the basis for the problem is widespread (for example, a natural disaster such as a hurricane or earthquake).

For example, fiber-optic cable and electric power lines often follow railroad rights-of-way, creating a geographic interdependency among the transportation, communications, and electric power infrastructure. Consequently, a train derailment could disrupt the entire infrastructure within that geographic area at the same time. This, in turn, would affect the railway company's ability to communicate during repairs, as well as the responsible electric company's ability to restore service.

Identifying, understanding, and analyzing the interdependencies among infrastructures have assumed increasing importance during the past 20 years. Key regulatory, technological, and economic changes have significantly altered the relationships among infrastructure, and the advances in information technology have led to substantially more interconnected and complex infrastructures. The infrastructure owner and regulatory agencies have accepted the importance of infrastructure interdependencies and the need to understand more fully their influences on infrastructure operations and behavior. A detailed example of the application of the interdependency model is provided in Chapter 4.

1.5 Understanding Roles, Responsibilities, and Community Engagement

The roles and responsibilities of both emergency managers and emergency responders are defined in significant detail. These two groups cannot be successful without willing participation from business/industry and the general public. These two groups must actively engage each other through the hazard/threat planning and mitigation process and may assist each other during response and recovery phases.

Geospatial technologies play an increasingly important role in facilitating the requisite level of community engagement by sharing information about plans and mitigating actions in a highly available and easy-to-understand format: the electronic map. In fact, the advent of crowd sourcing, whereby citizens and businesses may map items of concern or provide reports from the field as an event unfolds, provides valuable two-way communication between the emergency management and response community and those who are served by them.

The transition of this communication from verbal (radio, telephone) or text-based reports to a map represents a tremendous improvement in the speed and effectiveness of communication pertaining to disasters and critical infrastructure protection. What remains is a need to understand the consequences of such geospatially based communication and how it may best be employed.

References

BBC News. (2008). China anger over "shoddy schools." BBC News Online, May 15. http://news.bbc.co.uk/2/hi/asia-pacific/7400524.stm.

Clinton, W.J. (1998). Subject: Critical infrastructure protection. Presidential Decision Directive/NSC 63. May 22.

Federal Emergency Management Agency. (2008). ESF annexes introduction.

Federal Emergency Management Agency. (2013). IS-100.B: Introduction to incident command system, ICS-100.

Fulmer, J.E. (2009). What in the world is infrastructure? *Infrastructure Investor*, July/August, pp. 30–32.

Geospatial Information and Technology Association. (2008). Birds of a Feather Committee's report on critical infrastructure interdependencies.

Guha-Sapir, D., Hargitt, D., and Hoyois, O. (2004). *Thirty Years of Natural Disasters 1974–2003: The Numbers*. Centre for Research on the Epidemiology of Disasters, Presses Universitaires de Luvain, Brussels.

Hazus. http://www.fema.gov/hazus.

Hennessy-Fiske, M. (2011). Japan earthquake: Insurance cost for quake alone pegged at $35 billion, AIR says. *Los Angeles Times*, March 13. http://articles.latimes.com/2011/mar/13/world/la-fgw-japan-quake-insurance-20110314.

Hoyois, P., Below, R., Scheuren, J.-M., and Guha-Sapir, D. (2007). *Annual Disaster Statistical Review: Numbers and Trends 2006*. Centre for Research on the Epidemiology of Disasters, Presses Universitaires de Luvain, Brussels.

International Charter. (2013a). Home page. http://www.disasterscharter.org/web/charter/home.

International Charter. (2013b). Charter activations. http://www.disastercharter. org/web/charter/activations.

Kim, V. (2011). Japan damage could reach $235 billion, World Bank estimates. *Los Angeles Times*, March 21. http://www.latimes.com/business/la-fgw-japan-quake-world-bank-20110322,0,3799976.story.

Long, T. (2008). Dec. 3, 1984: Bhopal, "Worst industrial accident in history." *Wired*, December. http://www.wired.com/science/discoveries/news/2008/12/dayintech_1203.

National Police Agency of Japan. (2013). Damage situation and police countermeasures associated with 2011 Tohoku district—Off the Pacific Ocean Earthquake. Countermeasures for the Great East Japan Earthquake, April 10. http://www. npa.go.jp/archive/keibi/biki/higaijokyo_e.pdf.

Presidential Decision Directive 63. (1998). Critical Infrastructure Protection: Sector Coordinators. 63 FR 41804. http://www.ciao.gov.

President's Commission on Critical Infrastructure Protection. (1997). *Critical Foundations: Protecting America's Infrastructure: The Report of the President's Commission on Critical Infrastructure Protection*. The Commission, Washington, DC. http://www.ciao.gov.

Rinaldi, S.M., Peerenboom, J.P., and Kelly, T.K. (2001). Identifying, understanding, and analyzing critical infrastructure interdependencies. *IEEE Control Systems*, 21(6), 11–25.

2

Basics of Geographic Information Systems

2.1 The Purpose of Maps

The eighteenth-century philosopher Immanuel Kant taught that people could acquire understanding through three distinct viewpoints: the perspective of formal logic (mathematics), the perspective of time (history), and the perspective of space (geography). This last perspective emphasizes the importance of distance, site characteristics, and relative location in describing the relationships among several objects or facilities.

It is the perspective of space for which human beings are functionally designed. Two eyes facing forward enable depth perception. The symmetrical positioning of ears at the side of the head enables stereo-location of the source of a noise. Finally, the nervous system provides humans with *proprioception*, or a sense of how our body parts are positioned in space.

Moreover, human brains are wired inherently to perceive our world from a spatial perspective. Maps are the primary means of representing such relationships. Maps are analytical tools that depict spatial relationships and portray objects from the perspective of space. The power of maps rests on their synoptic (two-dimensional) representation of complex phenomena.

Because humans have an ability to more readily comprehend information presented from such spatially based perspectives, maps more closely resemble how we, as a species, interpret and understand our world (Sachs, 2010). To paraphrase the ancient wisdom, "a map is worth a thousand words."

For thousands of years, maps have been used for navigation, for political ambition, and for storing information of a spatial nature. As humankind's knowledge of the world increased, maps grew increasingly complex. Every map is a storehouse of information, but there is a physical limit to the amount of information that can be represented.

To maintain the usefulness of maps, cartographers (map makers) found it necessary to simplify maps by presenting small sets of information. One map in a set might display roads, while another might display agricultural production. When necessary, a special purpose map could be drawn by selecting from the information found on two or more maps.

More involved procedures for integrating spatial data into cartographic products, likely developed during Napoleon's campaigns in the eighteenth century, involved the layering of geographic data. Individual map themes, termed layers, were drawn upon hinged sheets of glass and added or removed from atop a base map depicting the locations of key features common to all themes.

The United States may owe its very existence to such an approach. Louis-Alexandre Berthier created maps depicting the location of British troop movements for Rochambeau during the Battle of Yorktown, thus enabling successful cannonade fire upon the enemy positioned in enfilade, a military formation's exposure to enemy fire along its longest axis (Rice and Brown, 1972).

As is common with modern spatial technologies, such techniques frequently crossed disciplines. The most famously known example is that of Dr. John Snow employing the approach during the 1854 epidemic of cholera in London, thereby enabling him to geo-locate the contaminated well at Broad Street as the source of infection (Snow, 1855).

However, this approach to analysis can be cumbersome and, depending on the complexity of information, subject to error. Until the introduction of digital computers and cartographers, no mechanism was available for fast, effective information extraction from these storehouses of knowledge.

Much of the work performed manually by cartographers, including actually drawing lines or patterns on a map and casting a map projection, can be done more accurately and more rapidly using computers. This method of map production is known formally as *computer-assisted cartography*, which recognizes the role still played by cartographers in such tasks as map design, layout, feature placement, and generalization. However, developments in the area of artificial intelligence are gradually altering the proportion of human involvement, and many professionals now refer to this method of production simply as *computer cartography*.

Because modern computers are digital in nature and require digital representations of maps for processing, the term *digital cartography* is gaining in popularity. The complexity of products is steadily moving from that of a static map, or one whereby a user may turn thematic overlays on or off, to those that are displayed from a three-dimensional perspective or incorporate animation as a means of depicting a fourth dimension (time).

The maps produced from such databases are called *digital maps*. As demonstrated, digital maps and computerized databases (which may be used and queried by several people simultaneously) are of immense value to engineers, comptrollers, planners, and managers. To extend the metaphor, the combination of a digital map and database is worth a thousand "megawords."

Although there were earlier development efforts, cartographers began to use computers in civilian production environments in the 1960s. By the early 1970s, efforts were underway in a few federal agencies, but according to a report by the Office of Management and Budget (1973, p. 158), there was little coordination, and "very few modern computer/microfilm-assisted data-handling systems [were] in use." In partial response to such criticisms, several federal agencies worked to integrate geographic information systems (GIS) into their workflows. Just 5 years later, Robinson et al. (1978, p. 4) were able to report:

> Recently it has become common to convert spatial phenomena to digital form and store the data on tapes or discs. These data can then be manipulated by a computer to supply answers to questions that formerly required a drawn map.... This stored geographic information is referred to as a [database].

The most notable effort was that of the U.S. Bureau of the Census and headed by James Corbett during the 1970s. His work focused on the translation of geographies important to the census process—hence the delineation of congressional districts and the accompanying representation foundational to the U.S. form of democratic governance. This goal was achieved through the creation of a digital format called Dual Independent Map Encoding (DIME) (Corbett, 1979).

The DIME approach extended the concepts of topology, or the digital construct whereby the spatial relationships among objects are maintained, an approach first proposed by Tomlinson in CanadaGIS (Tomlinson et al., 1976). This and subsequent related efforts, such as the creation of the Topologically Integrated Geographic Encoding and Referencing (TIGER) system (Broome and Meixler, 1990), created a significant new marketplace for geographic information systems software and solutions, and paved the way for commercial success by companies such as Esri.

As noted earlier, Kant taught that the perspective of space is fundamental to human thought. This perspective is validated as the complexity of cartography—the art and science of map making—has grown from its beginnings on a stone tablet 14,000 years ago (Govan, 2009) to the increasingly sophisticated computer mapping systems of today, referred to as GIS. GIS represents the culmination of 14 millennia of spatial analysis and the development of a powerful analytical tool.

2.2 Overview of Technology

Computer graphics in general and computer cartography in particular were a natural outgrowth of the computer revolution, expressing the human need to represent data in a graphic summary form to aid understanding. As with other forms of computer use, computer cartography is performed using hardware (physical equipment), software (the methods of instructing the computer hardware how to perform necessary tasks), and "human ware" (the decisions, questions, and design choices made by professional system designers).

Computer hardware consists of five classes of components: processors, input devices, output devices, storage facilities, and communication devices. The earliest computer mapping systems used mainframe computers as the system processor. By the 1960s, mainframes were being replaced by minicomputers, which in turn gave way to graphic workstations and microcomputers.

Although the systems were becoming smaller, they also were becoming computationally more powerful and substantially less expensive. A typical minicomputer-based computer mapping system of the mid-1980s with one processor, four interactive graphics workstations, storage, and an output device featuring appropriate mapping software carried a price tag approaching $1 million. Ten years later, the same (or better) functionality was available for less than $100,000.

Ten years into the twenty-first century, the price tag for an even more powerful single-user system had been reduced to $5,000 or less. Furthermore, the advent of cloud computing has shifted much of the computational processing from the local computer, allowing many tasks to be moved to smart phones and tablets.

Similar changes—increased power, smaller size, and reduced cost—also characterized other hardware components. The earliest display devices for computer graphics were similar to electric typewriters in design and output. In the 1950s, maps and other drawings were produced using a line printer and standard alphanumeric keystrokes. Such a method permitted five or six shades and patterns to be represented and distinguished in addition to the rough outlines of the regions being shaded.

By the 1960s, printers with overstrike capability were widely available. This capability increased the number of visually distinguishable shades and patterns to 10 or more. In addition, cathode ray tube (CRT) devices similar to early televisions were introduced. Pen plotting devices, laser printers, three-dimensional pneumatic tabletop displays, and other devices evolved to present the full range of cartographic output. Again, the advent of cloud computing has enabled output, in the form of maps and route displays, to be presented on a broad range of portable devices.

Graphic data input devices evolved in a similar fashion. The development of these devices evolved into two distinct ways of representing graphics: vector representations and raster representations.

The work of a draftsman using a straightedge and pencil to produce architectural drawings is an example of a vector representation. The digital version of vector graphics consists of points, lines, and polygons drawn using the digital equivalent of a pencil.

Originally, the bulk of vector digitizing was performed using manual digitizing devices. The arm and beam devices used in the 1960s and 1970s were replaced in the 1980s by digitizing tablets that used electromagnetic, electrostatic, or sonic cursors to record locations. These, in turn, gave way in the 1990s to precision digitizing cursors and computer mice.

In simple terms, graphic information is entered by first defining a coordinate system and then recording X, Y coordinates for points on a line using a mouse. The locations of any two points define a line segment; by convention, these objects are referred to as endpoints and line strings, or as arcs and nodes. A technician enters complex shapes by capturing the X, Y coordinates for every vertex or point of inflection of a feature (for example, a building footprint).

For complex shapes, such as elevation contours, the cursor may be set to record the locations of a continuous stream of points. After digitizing is complete, this large set of points is thinned (reduced in size) to eliminate redundant or superfluous points.

Raster graphics are significantly different in structure and methods of capture. For many years, artists and cartographers performed changes in scale using a grid overlay technique. First, a grid was superimposed on the original drawing. Then, a grid with larger cells was superimposed on a clean drawing service. Finally, the contents of the original drawing were transferred cell by cell to the new surface. In other words, the drawing was broken into discrete components to render a complex image in a more easily manipulated form.

Analogously, raster scanning breaks down a complex image by recording information (for example, color, shade, hue, tone, or electromagnetic spectral reflectance) about components of that image. In a sense, this is a taxonomic exercise because a continuous surface is rendered in discrete packets of information. This characteristic implies that statistical procedures, such as nearest-neighbor, principal components, and multidimensional scaling statistics, may be, and indeed are, used to classify the information.

As van Dam (1984, p. 149) noted, a raster graphic is an electronic version of "the pointillist technique developed by the 19th century French Impressionist painter Georges Seurat." This type of graphic representation is captured by means of scanning, with the resolution of the scan determining the size of the "points" in the image.

Two methods of raster data encoding are now in common use. The first of these is the automatic recording and entry of raster data at the moment of initial capture. An example is the digital capture of images by the Landsat remote sensing platform, a satellite that captures discrete picture elements, or pixels, of a given size.

The Landsat multispectral scanning system targets specific bands of the electromagnetic spectrum. The spectral reflectance (in-band radiance) in the selected bands is recorded by pixel and used to identify a spectral signature. These signatures are compiled, analyzed, and used to build a composite image of a region. The resulting data set is called a tessellation model.

The second method of encoding is automated raster scanning of existing source documents. Raster scanning is the rectangular pattern of image storage and transmission used in the majority of computer image systems. It is a systematic process of covering the area progressively, one line at a time.

Some systems record binary data representing the points, lines, and polygons on a map. Pixels that are black are coded as 1s, and pixels that are white are coded as 0s. Other more sophisticated systems scan and report colors or grayscale intensities on the source documents.

The last two classes of hardware—storage facilities and communication devices—have not only evolved to become faster, better, and cheaper, but in many cases have also converged. The explosive development of cloud computing infrastructure as a service (IaaS), software as a service (SaaS), and platform as a service (PaaS) offerings has had a profound effect on the development of computer mapping systems in the twenty-first century.

2.3 Parallel Origins in Automated Cartography and Planning

The primary concern in early computer mapping was with the production of finished paper maps that recreated the experience of traditional, hand-drawn maps. Over time, this goal was superseded by a desire for visual display of data in digital environments (for example, on in-car navigation screens). However, the origins of the technology continued to shape its evolution for several decades.

Two distinctly different approaches to computer-assisted mapping were taken by early industry developers. The first approach, exemplified by firms such as Intergraph and Autodesk, focused on recreating the traditional manual methods of map production.

Traditionally, maps were produced using multiple printing plates called *map separates*. Each separate was used to print one color of the map. For example, one separate would be designed to print all features that would be colored blue on the map. This blue separate would be used to print hydrology: rivers, streams, lakes, seas, and oceans. Similarly, a green separate would be used to show woodland features, and a brown separate would be used to print contour lines depicting topography.

These separates were produced at the same size as the finished map using dimensionally stable drafting materials. The separates, which were produced as photographic negatives, were used to produce printing plates. The printing plates were registered using a pin and slot system that ensured exact positioning of the plates during the printing process. After each plate was used to print a color, the ink was allowed to dry and the next layer of ink was printed. Thus, the separates were often referred to as *layers*, a term that was carried into the language of digital cartography (Figure 2.1).

The layer approach to computer-assisted mapping was brought to life by technology firms that had grown within the traditions of the engineering

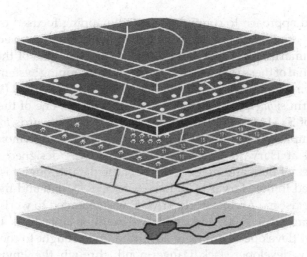

FIGURE 2.1
Layer approach to map creation.

community. Typically, the maps produced for engineering, such as utility company plans and highway construction drawings, were produced at large scales.

Common map scales for an engineering project might be 1:400 (1 cm on the map representing 400 cm in the real world) for base maps, 1:100 or 1:200 for construction diagrams, and 1:25 or 1:50 for detail drawings. Drawn at these scales, maps are particularly useful for construction and maintenance.

Over time, this approach to computer-assisted mapping evolved. Rather than drawing design notes and specifications on the maps and plans, designers began to place such information in a database that was "attached" to the drawings. For example, every telephone pole or electric power pole is identified with a particular identification (ID) number. Pole number 61,655, for example, possesses certain characteristics or attributes, including height, diameter, and date of placement in service. Traditionally, this information would be drawn on the map and viewed by engineers as and when needed.

GIS developers quickly realized the benefits of placing such information in a database. Data stored in a database could be searched by computer and used for numerous other purposes. For example, financial analysts could perform tasks such as asset depreciation and tax allocation studies. Similarly, planners could examine the age of equipment and schedule replacement as necessary.

The transition from pure computer graphics to a linked system of graphics and databases represented one of the most significant advances in this approach to computer mapping. The ability to design and build infrastructure was enhanced. Equally importantly, the reach of such systems was expanded to incorporate large-area planning based on access to large data sets of entire companies and multistate utilities.

The second approach to computer-assisted mapping focused on the integration of multiple data sets from its outset, with graphics used—at least initially—primarily for display and visual inspection. One of the first and most influential of these tools was the Synagraphic Mapping System (SYMAP) digital mapping tools and the related SYMVU map viewing tool (Waldheim, 2011). The term *synagraphic*, created by Howard Fisher, one of the principal developers of SYMAP, combines the Greek roots of the words *together* and *graphics* to emphasize the idea of *seeing things together* (Dangermond, 2014).

Developed at Harvard in the 1960s, SYMAP was designed to look at large volumes of data and make sense of those data through visual analysis. Developed in the Harvard Graduate School of Design and the Harvard School of Landscape Architecture, SYMAP was intended to work at smaller scales (for example, 1:50,000 and smaller) to examine large areas. Ultimately, the tools first developed here were advanced and brought to the public by another early developer, Jack Dangermond, through the Environmental Systems Research Institute, now known as Esri.

One approach—the data-driven approach—focused on automated mapping with the subsequent addition of databases. The second approach—the graphics-driven approach—focused on data analysis with the collateral display of graphic results to facilitate understanding. From the 1970s through the 1990s, the two approaches grew toward each other at a rapid pace.

For example, differences in the scale of use began to disappear as the database capabilities of automated mapping tools improved and as data-driven approaches integrated increasingly sophisticated graphics and provided methods of accessing external graphics files. Other changes affected this transition as well, but nothing more than the evolution of databases and database management systems.

2.4 Evolution as a Data-Driven Fundamental Information Technology

2.4.1 Introduction

At one time, computer users were expected to be computer programmers as well. This meant people who wished to solve a problem using a computer needed to learn a computer language (more correctly, a logical computer code) designed for communication between humans and machines. Among the most common languages were FORTRAN (Formula Translation), COBOL (Common Business-Oriented Language), and later PL1 (Programming Language-1). It also meant computer users frequently duplicated the efforts of others because many tasks assigned to computers were identical or quite similar.

To resolve the problem of duplication and make time- and cost-efficient programs for specific tasks available to a larger group of users, several libraries of programs were developed, such as the International Mathematics Subroutine Library (IMSL). The concept of a library ultimately was extended and packages of programs were made available in a coherent form, which made access to the computer much easier. The development and diffusion of these packages has made it possible for people with comparatively little training or interest in computers to use computers for many different purposes.

As the volume of digital data began to grow, it became obvious to the profession that simple graphic display would not be sufficient to support the evolution of computer mapping. Programmers began to focus their attention on the problem of how to store and access large amounts of data efficiently. As noted by Blasgen (1982),

> Computer systems are increasingly used to aid in the management of information, and as a result, new kinds of data-oriented software and hardware are needed to enhance the ability of the computer to carry out this task. [Database systems] are computer systems devoted to the management of relatively persistent data. The computer software employed in a database system is called a database management system (DBMS).

The software used with a database is known generically as a database management system (DBMS). Such a system generally will have provisions for data structure definition as well as for database creation, maintenance, query, reporting, and verification. Blasgen (1982) observed that in 1981, an estimated 50 companies were marketing 54 different DBMS packages.

2.4.2 Database Structures

Of the several methods of classifying databases used by software engineers, the most important distinction is that of database structure. There are several major types of database structures that reflect the purpose for which they were constructed and the uses to which they are best suited. Two fundamental concepts remain invariant requirements for the use of database technologies within the realm of geospatial technologies.

First, and perhaps foremost, is that all database models must accommodate information about a geographic object as well as its location. This can be accomplished through topological (adjacency) descriptions that may be correctly interpreted and mapped or through direct encoding of coordinate information.

The second concept is the notion of data independence whereby the arrangement and storage of information are separate, often physically, from the GIS application. This separation enables changes to other elements of the system, such as the color or thickness of a line, to be maintained without performing an underlying state change to the data affected. The concept enables both the ability to consume or act upon a single data set using multiple software applications and the ability to render multiple versions of the underlying database simultaneously.

For example, the latter traits are often desirable when multiple systems functions (for example, feature edit and cartographic display) are working with the same data simultaneously at two geographically distinct locations within an engineering firm. Both are essential to a fundamental tenet for managing geospatial data: the ideal GIS database is one that maximizes the uniqueness of every feature while minimizing the total data quantity.

A sequential database consists of data whose sequence (or order of collection, storage, and retrieval) constitutes a critical component of the information contained by those data. This is the most simple of geospatial data structures and relates most easily to the concept of topology, or the rules governing the spatial arrangement of geographic features. Another way of phrasing this is to observe that some data make sense only in a particular sequence (for example, alphabetical or chronological). One example relevant to cartography is navigation.

For the early navigator who ventured beyond the sight of land, one solution to the problem of defining direction was the use of *portolan* charts. These port-finding charts of the fourteenth and fifteenth centuries showed a series of parallel lines indicating north–south location and one or more compass roses. The chart was covered with crisscrossing lines emanating from these roses, lines of constant compass bearing known as *rhumb lines*. Sailors proceeded from port to port using a method of navigation termed *dead reckoning*. The collection of directions for turns and bearings constituted a sequential database.

A second example of sequential databases is perhaps the oldest survey system—the metes-and-bounds system of surveying. In this survey system, a specified location (usually a substantial physical landmark) is used as the initial point of origin. From this point, the direction and distance to each succeeding characteristic point are recorded. The direction is specified as an angle, in terms of compass-identified north (also known as a north azimuth).

The survey continues, as each successive endpoint becomes, in turn, an origin point. From the last endpoint, the transit is "closed" by returning to the first origin point. Any closure error (that is, any cumulative error resulting from field measurement errors or rounding-off calculations) is corrected.

Much surveying in early U.S. history was performed using metes-and-bounds procedures. Even in the present day, many private home lots and property titles are specified by metes and bounds. Because surveys are performed from point to point in a sequence, the order in which data are collected carries valuable information. Therefore, the most common survey database structures store the sequence of observations, as well as their distance and direction values (Figures 2.2 and 2.3).

In the future, with GPS observations of increasing importance for surveys, the sequence of observations will become less significant. In the near term, the sequential database structure predominates.

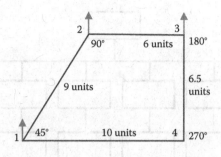

FIGURE 2.2
Sequential data example.

Record Number	Start Point	End Point	Angle	Distance
1	1	2	45	9
2	2	3	90	6
3	3	4	180	6.5
4	4	1 (5)	270	10

FIGURE 2.3
Sequential database table.

A hierarchical database structure organizes data in parent–child relationships. The highest-level parent (for example, great-great-grandparent) is the fundamental work unit (for example, a project or a company). This parent will have several children, which in turn have children.

A good analogy to a hierarchical database is a tree trunk with its branches, each branch giving rise to smaller branches. For this reason, this database structure is also termed a tree-like or *dendritic* structure.

To clarify this discussion, consider Figure 2.4, which presents an example from the telephone industry. Each of the numbered items is called an *entity*. An entity is similar to a classification, such as human being, dog, or vehicle. An entity occurrence is an actual value. More generally, programmers call this a *record*.

For the entities listed, example entity occurrences are George Washington, Spot, and automobile. Every entity has characteristics called *attributes*. Human beings have the attributes of height, weight, age, sex, and so forth. An attribute occurrence is an actual value. Consider the entity occurrence automobile. For the attributes year, transmission, color, model, options, and brand, some typical attribute occurrences might be 1970, 5-speed, red, convertible, magnesium-wheeled, and Corvette.

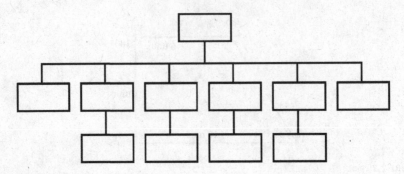

FIGURE 2.4
Hierarchical database structure.

The representation in Figure 2.4 is termed a *database schema*. Databases may be described more precisely using a data definition language (DDL). In the early days of the U.S. Postal Service, mailmen used compartments called pigeonholes to sort mail. A DDL tells programmers which pigeonholes to use for entity occurrences and attribute occurrences. For the sample shown in Figure 2.4, there will be a row of pigeonholes for the entity street. The length of that row will be determined by the number of actual streets in the company's plant location record (PLR) or similar asset management source document (that is, the number of entity occurrences).

To locate a particular entity in a hierarchical database, the DBMS searches down through the structure. It is important to note that the encompassing database key is reliant upon parent–child relationships constrained to one-to-one or one-to-many relationships. Such relationships are termed cardinality. For this reason, the ways in which the data will be accessed and manipulated must be defined precisely in advance.

This factor means that searches are comparatively fast and data access is very efficient. However, modifications to the structure are accomplished only with great difficulty. If the relationships among data change frequently, a hierarchical database structure will require substantial programming effort to maintain operations. Moreover, data redundancy may be unreasonably great.

A network database structure is similar in many respects to a hierarchical structure. The major difference is that whereas a hierarchical structural cardinality is defined by many-to-one parent–child relationships, a network structure is defined by many-to-many multiple-parent relationships (Figure 2.5). This structure permits reasonably high-speed data access, while reducing data redundancy.

The key structural feature of a network database structure is the use of pointers in the data records to indicate cross-connections. These cross-connections add complexity to the task of accessing data. The more complex the access structure, the more time-consuming the access. Alternations in the chain of connections may adversely affect system performance.

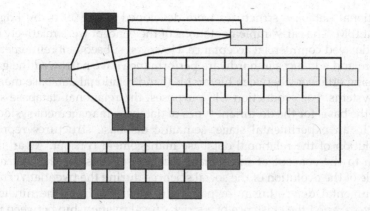

FIGURE 2.5
Network database structure.

For this reason, the applications should be defined as well as possible during the initial database design. The network data model, as defined by the Conference on Data Systems Languages (CODASYL), is also largely responsible for the ability to create a data audit trail and the accompanying creation of database administrators.

In a relational database structure, data are stored in tables (relations) of rows (elements) and columns (attributes) (Figure 2.6). To access the data in these stored tables, applications called virtual tables (views) are defined as needed. Only a description of the virtual table is stored in the computer, so they exist only when accessed. Consequently, data redundancy is eliminated. However, the ability to create virtual tables requires the use of keys that serve to uniquely identify and retrieve information from the database. Loss or corruption of these keys or the failure to ensure key uniqueness renders the underlying database useless.

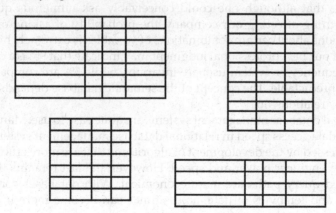

FIGURE 2.6
Relational database structure.

Relational database structures were developed in the 1970s by Edgar F. Codd at IBM. The early implementations of the concept were relatively slow, which delayed commercial acceptance. However, subsequent refinements in software and advances in hardware performance have narrowed the gap in processing efficiency between hierarchical and relational database management systems. For production GIS purposes, the relational database structure is the basis for the dominant types of database management systems.

Still in an experimental stage, semantic database structures represent an evolution of the relational database management concept. As an introduction to the concept of the semantic database, consider the instructive example of the evolution of the social sciences during the twentieth century. Environmental determinism—a paradigm inherited from the nineteenth century—argued the existence of a strict, causal relationship between natural environments and human cultures.

Possibilism, which evolved as a reaction to determinism, recorded and emphasized the diversity of responses to environmental stimuli and invoked other explanatory agents. Probabilism, which remains in favor in many circles, recognized that although human cultures might interact with and respond to the world around them in a variety of ways, certain types of development and modes of interaction were much more likely than others.

In an analogous manner, the network and hierarchical database models afford little flexibility in queries. After the environment (the DDL) is specified, the nature of the resulting culture (interaction with the system) is determined. One principal advantage of a relational database structure is the flexibility to support a wide range of queries, none of which need be defined before the fact. In this regard, the relational database model affords access to all possible combinations of variables and records.

Yet human thought is neither exhaustively, rigorously logical nor completely unconditioned by prior experience. We know from our training and experience that although one could conceivably ask numerous questions about anything affecting our company, the probability of asking only certain questions about certain combinations of variables is extremely high. The exercise of our best professional judgment may indicate that 95% of our queries (in frequency or system use) are invariant, while 5% are comparatively rare or unpredictable. The concept of the semantic database depends on this probability relationship.

Relational database management systems are quite easy to use. Many issues related to data access speed in relational database management systems have been addressed by the development of algorithms that are more efficient and computing engineer power and speed. However, the fact remains that predetermined query paths are more economical. With databases containing gigabytes and terabytes of data, access times can become more important than any other single operating parameter.

The semantic database attempts to reconcile these issues by predefining the most probable query paths (for system optimization) while retaining the

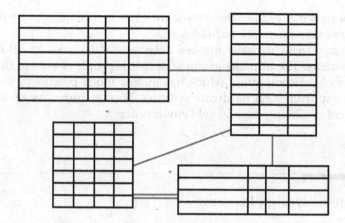

FIGURE 2.7
Semantic database structure.

flexibility offered by the relational database. In effect, the semantic database concept links multiple relational databases and tables based on the interaction of users with the system (Figure 2.7).

At least three distinct methodologies for implementing the semantic database are under investigation and development. One approach seeks to adapt new and existing relational models to accommodate *hard-coded* query paths. A second approach depends on the user-specific, customized development of rules and specifications that support emulation of ad hoc query capability in a network/hierarchical model environment.

The third, most challenging approach seeks to develop a fresh model that grows from our knowledge of artificial intelligence and the next generation. In some respects, the semantic database offers the functionality of the neural networks (systems capable of machine learning and pattern recognition due to their adaptive nature) first conceived in the 1950s.

Table 2.1 summarizes several key characteristics of these database structures. However, we emphasize the values of these characteristics are not absolute, but are relative to the other database structure types. For example,

TABLE 2.1

Characteristics of Database Structures

Structure	Time Frames	Schema Modification	Modeling Power	Application Development	System Overhead	Relative Performance
Hierarchical	Early 1970s	Difficult	Low	Difficult	Low	Medium
Network	Late 1970s	Difficult	Medium	Difficult	Low	Medium
Relational	1980s	Moderate	Medium	Easy	High	Low
Semantic	1990s	Easy	High	Easy	Medium	High

the performance of each of these structures has been enhanced by massive improvements in processor technologies.

These improvements have masked relative inefficiencies in all but the largest systems. The ultimate example of this principle is seen in the rapid growth of cloud computing, which has moved much processing and data storage away from local hardware systems. This, in turn, has enabled the deployment of GIS into mobile field environments.

2.5 Systems Design Process

An alternative view of a geographic information system is as an amalgamation of its constituent parts: hardware, software, data, people, and procedures. Hardware includes the required physical computers and data collection devices. Software includes the GIS package itself, as well as operating systems, data processing and database systems that produce or consume data, and other packages that must be used and integrated for a GIS to function correctly and reliably.

Data are all elements of information collected, stored, and used, and output from a GIS. Data are collected through a translation of observable or perceived phenomena into electronic format and must adhere to a structured syntax to allow for machine processing. Syntax is an example of the many procedures and well-defined and -documented processes required to use a GIS successfully. People are the most important part of any system, as they plan for, operate, and maintain all GIS and are the only component capable of repairing or remediating all other aspects.

All elements must function together in true systemic fashion to achieve the basic levels of functionality required of any GIS: to capture, store, query, analyze, and view data, and to produce an output product or service. These functions may be simplified and viewed within the context of a pipeline, as shown in Figure 2.8.

A principle challenge in delivering effective geospatial services is that the source of data (geographic features) and many downstream, often unconstrained and unintended, uses occur away from the office. This means that some intermediary method must be employed to gather geospatial data and transport them to a location where it may be entered into a GIS for processing and storage. Likewise, delivering an appropriate final product to an end user requires transport.

Given this challenge, geospatial technologies must manage the logistics of data inputs to align them with the required products and services. Both *Thinking about GIS* (Tomlinson, 2011) and the *Local GIS Development Guide* (Becker et al., 1994) provide lengthy treatises about the methodologies used to develop large GIS, and *Building a GIS* (Peters, 2012) provides a path forward

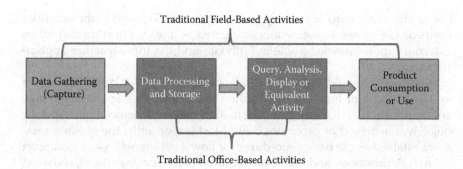

Traditional Field-Based Activities

| Data Gathering (Capture) | Data Processing and Storage | Query, Analysis, Display or Equivalent Activity | Product Consumption or Use |

Traditional Office-Based Activities

FIGURE 2.8
The GIS pipeline.

for systems architecture design strategies, albeit one best suited for an Esri-based system.

Senior levels of management must understand the value of geospatial technologies within an organization and commit their use with knowledge that a return on investment or improved efficiencies may take time to fulfill. Organizations must develop a positive attitude toward change prior to starting a GIS and should conduct educational seminars or briefings about the many uses of GIS whereby the advantages to both individual users and the organization are well documented and communicated. New technologies are often frightening to a workforce and may be seen as a means of replacing employees rather than empowering them.

Furthermore, organizations beginning GIS development should engage in a dialogue with the greater profession through professional societies, conferences, and meetings. An organization that is versed in the potential uses and benefits of spatial technologies and able to see how they may help achieve goals stands to gain more than those that do not.

Beyond the need for a qualified geospatial support team, the authors felt it was important to outline the GIS development life cycle for those considering the implementation of a GIS program to support emergency management, protect critical infrastructure, or provide any day-to-day operational need of an organization. The GIS development life cycle is a continual process and represents a long-term organizational commitment. Few, if any, GIS projects fail due to technical considerations.

Management decisions are the single most important component inherent in creating a successful GIS program. Management is able to understand the big-picture items, such as budget, user expectations, planning/operational/management needs, and similar processes that govern the overall success of a project. With this in mind, a team should be convened to guide the GIS development life cycle process through the following steps.

First, the team should perform a needs assessment. This is the most important part of the development process. All potential users should be identified, workflows documented, end products and services (GIS functions and the applications required to support them) identified, and responsibilities defined. The needs assessment is conducted as a team effort guided by an external expert knowledgeable in both GIS and the infrastructure application area.

The process should provide a guidance document that aptly provides a systematic look at how an organization views and uses spatial data, how communication and sharing of information may be enhanced, and specific objectives achieved or capacities built. Most importantly, the needs assessment establishes the basic procedures for how a system will work in concert with work processes, and define the required personnel, hardware, software, and data requirements.

Second, the team should create an implementation plan. The implementation plan provides a detailed look at how the system will be constructed and carries forward all of the fundamental aspects of data, hardware, software, procedures, and personnel defined during the needs assessment. It must contain well-defined milestones with supportive timelines and budgets.

The implementation plan should also explicitly state interdependencies and consequences that will help guide the overall development effort. It must contain detailed information about system and data architectures, the roles and expectations of all personnel involved, and testing and training plans that will ensure competency when the system goes live.

Third, the team must develop the theoretical model/framework for the system. This is the conceptual understanding of how work and its supporting data move within the proposed system. The theoretical model may be simulated to estimate hardware requirements, potential data or workflow bottlenecks, and similar challenges. Tools such as SLAM II or similar operations research/management simulation packages can be adapted to assist with especially large or complex systems.

Fourth, the team should conduct a survey of available data. Purchasing or collecting large quantities of new data is a frequent temptation and costly expense. Data requirements must be clearly stated with respect to both spatial and attribute precision and accuracy prior to any such effort.

Existing data resources that meet the project's needs are often available for free from open data sources such as Open Street Map or through a government-provided source, often reside in institutional records in paper format, or may be collected through the course of normal operations. Where required data do not exist, developers should consider data collection techniques such as LIDAR, which may facilitate feature extraction in a way that meets not only current but also potential future needs.

Fifth, the team should conduct a survey of GIS hardware and software. Organizations, particularly those with an engineering or emergency response aspect, often have multiple hardware system types and a wide variety of operating systems. While such enterprise-wide customization may permit greater flexibility to departments and individuals, it may prove a barrier to using some spatial technologies broadly or raise the longer-term cost of ownership and maintenance. Organizations are urged to consider not only immediate implementation costs for hardware and software, but also longer-term costs of ownership and the frequency and reliability of updates or evolutionary changes in platforms.

Sixth, the team should undertake detailed database planning and design. This step involves translation of the theoretical model into a logical model in support of the required applications. Entity relationship diagrams may be used and customized for spatial technology–specific tasks, such as the depiction of topology (the spatial relationship among objects). Organizations should choose a database design that maximizes the uniqueness (minimizes duplication) of data and which has optimal flexibility to support organizational growth and emergent needs. Detailed documentation is essential to database planning and design.

The seventh step is database construction. This aspect of systems development is not simply load and go, but rather entails detailed quality control and quality assurance plans. Care should be taken to consider the physical and electronic security of information.

The eighth step is applications development. While this process may begin in earnest earlier than indicated here, it may not be truly put into play until the final data model is constructed. Applications development must not only consider needs identified during the needs assessment process, but also, as with database construction, take care to provide appropriate access and security control. This is of special concern where a GIS may interface with supervisory control and data acquisition (SCADA) systems, personally identifiable information, or other sensitive data.

The ninth step is to conduct a pilot study and benchmark test. Pilot and benchmark testing should be conducted as expediently as possible and may accompany each phase of development. Successful results add momentum to projects and demonstrate worth early, providing an impetus for continued support of lengthy or expensive projects. When the tests fail, the information is particularly useful in correcting course before large, full-scale expenditures are made.

The tenth step is to review and modify the original plan, based on the results of the pilot study and benchmark tests. Information from each phase should be gathered and reexamined periodically, and especially before the next step, purchase of hardware and software, occurs. This often allows shifts to accommodate new versions or small changes in technical approach before committing large sums for purchasing GIS infrastructure.

Only after this careful planning and testing is complete can the next steps of full project deployment be undertaken with confidence.

1. **Acquisition of GIS hardware and software.** As with the purchase of a new car, software and hardware lose value and currency as soon as they are installed. Further time is lost if a significant time gap occurs between when these purchases occur and when they may

actually be put to practical use. Hardware and software purchases should be delayed as long as possible to prevent such loss in value.

2. **GIS integration.** Systems integration, after having been studied and tested extensively, is the penultimate step in systems development. To state the obvious, rollovers must be carefully planned and executed in a controlled environment only after sufficient disaster/recovery precautions are taken.

3. **System testing.** Development and unit testing should be conducted when all approved system components are constructed and unit tested in the development environment. System components are defined as all programs, databases, tables, procedures, policies, documentation, training materials, and test cases/data. The objective of this step is to complete all development activities so that the newly built system is ready for integration and system-level testing.

   The purpose of the quality assurance (QA) testing is to perform various system and QA functions to validate the integrity of the developed system. At the end of each testing phase, the results must be documented. Testing should be performed in the test environment. As part of testing, all proposed requirement changes must go through the formal change request and approval process. During this phase, initial user training is typically delivered along with the current working version of the application documentation.

4. **User training.** User acceptance testing verifies that the testing activities have been completed and have validated the business design of the newly developed or enhanced system. This includes reviewing the test results from the testing phase, and simulating the operational environment before going live with the system.

5. **GIS implementation.** During the actual implementation, two key activities are accomplished, including the migration from the existing system. This must be initiated, managed, verified, and completed. If there are any conversions, the implementation team must ensure the data have been properly transferred and the system is primed and ready for production processing, and ensure the cutover to the system takes place.

6. **GIS use and maintenance.** Metrics should be in place to measure and monitor the effectiveness of new systems and be used to evaluate how well the objectives defined in the needs assessment match reality. These metrics may be used to further refine and develop the system and provide an easier-to-understand cost basis for such decisions.

The GIS development life cycle described above is reasonably straightforward in most cases. As with all things, the process is substantially

complicated when applied to emergency management and response organizations. The latter require very scalable solutions that can rapidly expand
from peacetime operations to those required during a major disaster.

Staffing and GIS infrastructure recommendations must be prepared in a
scalable and modular format to meet the described roles. Further, applications needed during normal daily operations may not be required during
a crisis, and vice versa. One approach for contending with this duality is to
prepare two distinct needs assessments, one representative of each operational state, and then reconcile them into a common implementation plan
moving forward.

2.6 Going Mobile

One of the principle factors influencing the return on investment generated
from using a GIS is the speed at which the system may operate. The primary
bottlenecks in the GIS pipeline (Figure 2.8) have traditionally occurred where
data must be moved to a location external to the office environment. These
chokepoints most often occur when collecting data in the field or when dis-
tributing a resulting product or service to an end user.

Some of the most significant advances in GIS-based technologies are
represented in the transition from traditional survey-based data collection
methods in the field to the use of modern communications systems and data
transfer technologies. The use of such techniques is not without caution;
users should gain an appreciation of the traditional methods, as they are
informative and help drive the requirements process. Three case studies
are provided below that highlight the evolution of data capture methods and
their underlying requirements.

The U.S. Geological Survey (USGS) celebrated 125 years of topographic
mapping in 2009. The effort, begun by John Wesley Powell in 1884, represents
a massive undertaking that resulted in the production of approximately
33,000 7.5 × 7.5 ft map sheets drawn to a scale of 1:24,000. To date, this effort
represents the most comprehensive effort aimed at producing a true national
map for the United States.

Despite this lengthy effort, parts of Alaska are still only documented
at 1:50,000 scale. John Noble Wilford eloquently chronicles parts of this
endeavor through the experiences of Bradford Washington in the prologue
to his work *The Mapmakers* (2001):

> We pulled ourselves upright and tested the footing. Careful now, the
> pinnacle may be solid, but it is no more than two, three meters wide.
> And how barren! No vegetation sprouts from its hard, rust-colored
> mantle. No nest of eagles, no Anasazi shards, nothing. This place stood

out in defiance of the winds and floods that had shaped it and everything around as far as the eye could see. It stood resolute and solitary,
1,700 meters above sea level but reaching barely halfway to the Canyon
rim. Perfect for our purposes. We could see and be seen from all directions, from Hopi Point and Yaki, Yavapai Point and Cheops and Ra. Here
we could make more measurements needed to give the map its basic
frame of reference, its mathematical skeleton.

Early topographic maps were driven by a need to understand the lands
comprising the United States. To Powell, creating the topographic map
was basic science. He wrote, "I have long entertained the opinion that a
Government cannot do any scientific work of more value to the people at
large than by causing the construction of proper topographic maps of the
country" (Powell, 1885).

This need, in turn, drove the scale of the resulting map products, even-
tually providing the basis for the National Map Accuracy Standards (U.S.
Bureau of the Budget, 1947), where 90% of all well-defined features must be
depicted within 12.19 m of true location at a scale of 1:24,000. The example
stands as testament to use requirements (basic scientific inquiry) driving the
data collection process.

It is important to note the publisher of this standard: the Bureau of the
Budget. Increasing map accuracy and precision increases the cost of data
collection. Were these accuracy standards not arrived at as reasonable and
prudent for the end use of the topographic map product, costs might have
skyrocketed and a 125-year venture remained only partially completed to
this day. Regardless of the era in which data are collected, increasing precision and accuracy requirements by means disproportionate to end-user
requirements represents a substantial increase in investment and limitation
to return thereon.

Initial topographic map construction was done primarily through the use
of survey crews using theodolites (a precision instrument for measuring vertical and horizontal angles), survey chains (an obsolete unit of length dating
from the 1600s), and other now primitive surveying techniques in combination with large *survey crews* encompassing numerous technical field and
support staff and their provisioning in the field. Beyond being inconvenient
and expensive, this process was incredibly time-consuming. It could take
months, even years, for the data compiled by a survey crew to be translated
into a single published topographic map sheet.

Often, data were out of date before the map was ever published. Once
maps were published, their availability was limited, and it could again take
significant time to locate, purchase, and take delivery of a much needed map.
Thus, the quest to speed the data collection and product dissemination process was realized and fiscally incentivized.

Within the domain of critical infrastructure data collection, precision and
accuracy play a more dominant role. Infrastructure owners in general and

utilities in particular often struggle with the location of assets in the field and have traditionally attempted to collect location information with the greatest degree of precision and accuracy available at the time. This makes sense given the investment associated with building infrastructure systems and the fact that many elements reside below ground. In many cases, the age of installation and related data precision span decades or even hundreds of years.

Creating an infrastructure system typically starts with an engineering study to establish the requirements for the infrastructure elements. The selection of appropriate pipe or conductor sizes, the placement of lift pumps or compressor stations, and the size and location of a substation or switch house must be considered. Other details essential to an infrastructure system establish basic location-based requirements, such as potential routes or sites using small-scale maps (less accurate and precise than large-scale maps, but typically covering the type of large-scale geographic extents required for planning purposes).

A survey crew is then dispatched to create a large-scale map (higher accuracy and precision, smaller geographic extent) that may be used for design purposes. The resulting survey data are physically transferred from the survey equipment or logs and integrated as a base layer in a computer-aided design system.

The resulting product is passed to a construction team responsible for building the desired elements. A surveyor will often visit the construction site as work progresses to guide the build process geographically and to document changes made during the installation, such as realigning a pipeline to avoid a massive buried rock. The resulting modified construction map, now called an *as-built drawing*, with appropriate survey input, is brought to the office.

Data are then reconstructed from redlines (corrections to the original design drawing) through scanning and digitizing, downloaded from a survey instrument or logbook, or through similar methods. These data are used to modify a planned version of a GIS database, the results of which are transferred into an operational version of the GIS database. Note the ability of a relational database model to accommodate multiple states of the data, and that the increased need for precision and accuracy has lengthened the collection process and contributed substantially to costs.

Many infrastructure owners are now increasingly relying upon faster methods of data collection. For example, Salt River Project (SRP) went to the expense of installing a continuously operating reference system (CORS) in the early 2000s. This allowed SRP to perform a high-accuracy survey more quickly while ultimately decreasing equipment costs. SRP extended the idea a step further, realizing that much of the data it worked with needed to align with the data of partners.

SRP opened its CORS network for public access so that all mapping efforts in the area could be conducted using the same frame of reference. This and other similar benefits of collaboration are demonstrated by Leveson (2009). Again, end-user requirements drove the need for data precision and accuracy, and a transition to faster means of data collection in an effort to reduce production times and improve map quality in meaningful ways.

The advent of mobile data networks represents a significant leap forward in shortening the entire production life cycle. Connected data collection devices may send live data from the field for instant integration into a GIS database located in the cloud. Likewise, mobile devices may receive live feeds or services from a cloud-based GIS and use the transmitted data in the field for inquiry or analysis.

The latter represents a massive shift in where different elements of the GIS pipeline take place and, as new development efforts come to fruition, may represent the next massive opportunity for cost savings in GIS planning and development. Final products may therefore be generated on demand and serve far broader audiences of users due to a higher level of availability.

The 2010 earthquake that devastated portions of Haiti serves as a prime example of the rapid reshaping of the GIS pipeline. The most current and complete data available at the time of the earthquake were an incorrectly geo-referenced city map produced by the Defense Mapping Agency in the late 1960s. However, satellite imaging platforms were coincidently located overhead at the time of the quake, and GeoEye was able to capture relatively high-resolution imagery within hours of the shaking stopping.

These data were transferred to Google, and hundreds, if not thousands, of volunteers began digitizing in roadways, damaged buildings, and other features. These data were shared using Open Street Map, a crowd-sourcing map platform, and reflected the mapping of more than 100,000 uniquely identifiable features in just days, and with an underlying accuracy of a few meters. Simultaneously, geo-tagged text messages to help with rescue and recovery began to emerge from Haiti and were captured and mapped using another crowd-source data gathering platform called Ushahidi.

These and other data streams were integrated by the Center for Interdisciplinary Geospatial Information Technologies at Delta State University and converted into GeoPDF-based map sheets that could be used on mobile devices or printed to scale at letter size. The resulting map sheets were transferred electronically to the Harvard Center for Geographic Analysis, which hosted a distribution portal and other related services to support responders either already in the field or en route (Powell, 2010). In short, approximately 77,000 unique map sheets were created within 96 h and made readily available.

The potential for mobile data collection, especially using the crowd as data gatherers, the ability to process and analyze data in the field, and the ability to create or retrieve end products together represent the newest frontier in GIS development at the time of this writing. They also represent a significant challenge, as the underlying data quality, a primary driver in systems creation, may be lost or inappropriate for many of the purposes for which mobile technologies may be used. This and other related topics are explored further in subsequent chapters.

References

Blasgen, M.W. (1982). Data base systems. *Science*, February 12.

Becker, P., Calkins, H., Cote, C.J., Finneran, C., Hayes, G., and Murdoch, T. (1994). *GIS Development Guides*. New York State Archives and Records Administration, Albany.

Broome, F.R., and Meixler, D.B. (1990). The TIGER data base structure. *Cartography and Geographic Information Systems*, 17(1), 39–47.

Corbett, J.P. (1979). *United States Bureau of the Census Technical Paper 48: Topological Principles in Cartography*. U.S. Government Printing Office, Washington, DC.

Dangermond, J. (2014). Email communication to the authors. August 12.

Govan, F. (2009). World's oldest map: Spanish cave has landscape from 14,000 years ago. *The Telegraph*, August 6. http://www.telegraph.co.uk/news/worldnews/europe/spain/5978900/Worlds-oldest-map-Spanish-cave-has-landscape-from-14000-years-ago.html.

Leveson, I. (2009). *Socio-Economic Benefits Study: Scope the Value of CORS and GRAV-D*. Contract NCNL-0000-8-37007. Final report prepared for the National Geodetic Survey.

Office of Management and Budget. (1973). *Report of the Federal Mapping Task Force on Mapping, Charting, Geodesy and Surveying*. U.S. Government Printing Office, Washington, DC.

Peters, D. (2012). *Building a GIS: System Architecture Design Strategies for Managers*. Esri Press, Redlands, CA.

Powell, A. (2010). Portals into Haiti, Chile. *Harvard Gazette*. http://news.harvard.edu/gazette/story/2010/03/portals-into-haiti-chile/ (accessed October 14, 2014).

Powell, J.W. (1885). *Organization of Scientific Work of the General Government: Extracts from the Testimony Taken by the Joint Commission of the Senate and House of Representatives to Consider the Present Organizations of the Signal Service, Geological Survey, Coast and Geodetic Survey, and the Hydrographic Office of the Navy Department, with the View to Secure Greater Efficiency and Economy of Administrations*, 40. U.S. Government Printing Office, Washington, DC.

Rice, H.C., and Brown, A.S.K. (1972). *American Campaign of Rochambeau's Army, 1780, 1781, 1782, 1783: The Journals of Clermont-Crèvecour, Verger, and Berthier*. Princeton University Press, Princeton, NJ.

Robinson, A.H., Sale, R.D., and Morrison, J.L. (1978). *Elements of Cartography*, 4. 4th ed. John Wiley, New York.

Sachs, O. (2010). *The Mind's Eye*. Knopf, New York.

Snow, J. (1855). *On the Mode of Communication of Cholera*. 2nd ed. John Churchill, London.

Tomlinson, R.F. (2011). *Thinking about GIS*. 4th ed. Esri Press, Redlands, CA.

Tomlinson, R. (2014). *Thinking about GIS*. 5th ed. Esri Press, Redlands, CA.

Tomlinson, R.F., Calkins, H.W., and Marble, D.F. (1976). *Computer Handling of Geographic Data: An Examination of Selected Information Systems.* Natural Resources Research Report 13. Springer-Verlag, Paris.

U.S. Bureau of the Budget. (1947). *United States National Map Accuracy Standards.* Circular. Issued June 10, 1941, revised April 26, 1943, revised June 17, 1947.

Van Dam, A. (1984). Computer software for graphics. *Scientific American*, 251(3), 146–159.

Waldheim, C. (2011). The invention of GIS. *Harvard Gazette*, October 12. http://news.harvard.edu/gazette/story/2011/10/the-invention-of-gis/ (accessed August 18, 2014).

Wilford, J.N. (2001). *The Mapmakers.* Rev. ed. Vintage Books, New York.

3

Government's Application of GIS to CIP

3.1 ROADIC in Japan

The Road Administration Information Center (ROADIC) was originally created in 1986 as a result of several large-scale gas explosions that killed and injured hundreds of people and caused tremendous damage. These accidents were the result of a lack of knowledge of underground infrastructure that was encountered during excavation and construction activities. Given the nature of Japan's densely populated urban areas, most of the critical infrastructure lies beneath the roadways.

The gas line explosions and the need to coordinate road construction, coupled with available funding at the ministry level lent significant impetus to the formation of ROADIC. The Japanese national government saw the need to develop an approach to preserve public safety and improve response to accidents involving this significantly expanding public energy source. Consequently, it took the lead to organize ROADIC through its Ministry of Construction, Bureau of Roads, which enabled the foundation of the program in 1986. ROADIC was formally established as a nonprofit public entity.

A consortium of public and private members, ROADIC was set up as a national project to manage and protect the public utilities within the right-of-way. Following the initial implementation in metropolitan Tokyo in the mid-1980s, additional branches were established in 12 major urban centers across Japan. Cities include Tokyo, Sapporo, Chiba, Kawasaki, Kyoto, Yokohama, Nagoya, Osaka, Kobe, Hiroshima, Kitakyushu, and Fukuoka.

These branches coordinate with local government agencies and public utility companies, including electric, gas, sewer, water, subways, and communications. ROADIC is governed by a 20-member board of directors. The board members are largely local prefectural road administrators and utility representatives. The Ministry of Construction plays a fundamental role in the ROADIC program by coordinating with individual cities seeking to enter into the program.

ROADIC is organized into three major functions: general affairs, planning, and systems development. ROADIC maintains its own staff of about 80 people, many of whom previously worked at member organizations and have a significant amount of experience. In addition to a number of technical committees, a group of experts, consisting of one representative from each of ROADIC's 12 member centers, meets on a regular basis to discuss developments and organizational issues.

3.1.1 Economic Considerations and Benefits

The national government under the auspices of the Ministry of Construction provided significant initial funding to develop the program. The original cost of establishing the ROADIC program was approximately ¥9.5 billion, or $8.7 million, 60% of which was funded by the national government. The remainder was contributed by local governments and utilities companies.

Its annual operating budget is approximately ¥3.4 billion, or $3.1 million. The national government provides 50% of the annual operating budget. The balance is divided among the individual member organizations. Essentially, both taxpayers and ratepayers support ROADIC operations.

Several of the resulting benefits of the ROADIC program are associated with cost savings involving utility and construction coordination and management, and time reduction for road administrators managing the laborious permitting process.

For example, road administrators and utility companies can access maps and information on existing underground and aboveground infrastructure online from office and mobile computers that are linked to databases at each of the ROADIC branch locations. This system enables immediate access to utility and road data, planned designs of new utility facilities, and coordination of work schedules associated with construction and maintenance activities.

Specific asset management functions include the following:

- Renewal planning of assets and facilities
- Pipeline network analysis
- Design/provision for permit application
- Construction
- Data updating
- Data maintenance

3.1.2 Technology Platform

The Road Administration Information System (ROADIS) is a custom-developed computer mapping software application originally developed by Tokyo Gas as the Total Utility Mapping System (TUMSY). TUMSY supports a number of functions in the area of facility management, disaster management, infrastructure protection, and emergency operations.

ROADIS is based on a standard database format that is separated into a landbase (for example, transportation, planimetric, and terrain features) database and road utilities (for example, separate underground and above-ground utility features) database. This format ensures standard data exchange involving input and output of landbase and utilities data. In addition to utility infrastructure data, ROADIS includes data on the structure and building material of the underground infrastructure.

ROADIS is also integrated with a permitting system. All construction permits within the right-of-way are issued and managed centrally at the local level. This ensures proper coordination between all road and utility work.

ROADIC uses the concept of an expert user group to decide what new software functionality will be designed and implemented each year. This group also sets the priorities for the new development. The general plan is to review the entire system every 10 years, adding more branches and organizations along the way.

3.1.3 Interorganizational Relationships

As mentioned earlier, ROADIC coordinates with local government agencies and public utility companies, including electric, gas, water, sewer, trains, subways, and communications owners and operators. Members of ROADIC enter into a contractual agreement, similar to a proprietary confidentiality agreement, with ROADIC. Other coordinating agencies include the national highway offices, under the Ministry of Land, Infrastructure and Transport. These federal offices have jurisdiction over certain road facilities and infrastructure in the participating cities.

ROADIC serves as the focal point for all permit requests. All members submit preconstruction drawings and designs before work begins. Nonemergency work is planned well in advance and is fully coordinated to minimize traffic disruptions and unnecessary cutting of road pavement.

3.1.4 Standards

Standards were not a major issue for ROADIC because the road and utility data already used a standard format. TUMSY, the software application built as the original system for Tokyo Gas, provided the data later adopted by ROADIC. In effect, it became the default standard.

This is a sensitive issue of business culture. Most of the North American geographic information systems (GIS) community agrees that it is inappropriate to allow one organization or vendor to dictate data standards. Rather, the North American industry has spent a significant effort on developing—cooperatively—common data standards and system interoperability protocols.

Additionally, not all public utility companies in Japan use TUMSY. For example, Nippon Telegraph and Telephone (NTT), the Japanese national telecommunications company, uses its own system.

Communications between ROADIC and its member organizations are coordinated to ensure that feature classes are consistent. This process has evolved over the years to become routine operation. Because all members use the same landbase, individual organizations add their respective infrastructure and facilities in the common spatial database environment. Specifications assist with the translation of infrastructure and facilities data.

3.1.5 Critical Infrastructure Protection

The Japanese government seems to have less of a concern for protection against terrorism than the U.S. government. The nation's approach to critical infrastructure protection is driven more by concerns with natural phenomena, such as earthquakes, tsunami, floods, and volcanoes. At this point, ROADIC's primary purpose is to provide for road maintenance, construction, and management of infrastructure and facilities within the rights-of-way.

ROADIC is enabling government agencies and public utility companies, through the use of GIS, to increase coordination and sharing of vital infrastructure and facility information to support disaster planning and recovery activities, such as the powerful 2011 Tohoku earthquake and tsunami. That event resulted in 5,883 deaths, 6,150 injured people, and 2,651 people missing across 20 regions, as well as 129,225 buildings collapsed and more than 900,000 buildings damaged.

The earthquake and tsunami also caused extensive and severe structural damage in northeastern Japan, including heavy damage to roads and railways, as well as fires in many areas, a dam collapse, and the Fukushima Daiichi nuclear disaster. Approximately 4.4 million households in northeastern Japan were left without electricity and 1.5 million households were left without water.

In addition, Tokyo Gas is using TUMSY to coordinate emergency response with each city. The system connects the main control room at Tokyo Gas to computers installed in emergency vehicles for purposes of dispatch of service crews and emergency responders, retrieval of data, support of field operations, and coordination with city departments and other utility companies.

3.1.6 Return on Investment versus Public Safety

ROADIC officials have indicated that return on investment (ROI) or any cost benefit was not the business driver when the program was originally conceived. The business driver was crisis management at the time the ROADIC program was initiated. Public safety played a very significant role in the initial development period of ROADIC.

The fact that hundreds of people were killed and injured as a result of a set of major gas explosions required the national government of Japan to take action to ensure the future safety of its citizens. This was the impetus for the creation of the ROADIC program and ROADIS.

3.1.7 Coordinated Activities

There are a number of ongoing coordinate activities involving research, planning, construction, and maintenance between federal and local government agencies and private utility companies. These coordination activities include the following:

- Research on the use of road space and a system for more efficient use of this space
- Research on administration systems to manage roads and infrastructure to keep up with increasing demands of the population
- Proliferation of new technologies and standardization of the road and utility management systems
- Collection, analyses, and distribution of the latest data on roads and infrastructure
- Submission of applications for road occupancy permits
- Coordination of road work schedules
- Administration and protection of roads and infrastructure

3.1.8 Risk Management and Liability

Because ROADIS is a *closed* system, data and system access are limited to those organizations that are members. Restricting access to data and the system is enforced by a membership agreement with each member. This agreement includes strict system and data security policies. There is a process set up for new organizations interested in becoming a member.

3.1.9 Enhanced Effectiveness

Improved effectiveness has been a direct result of the development of a common landbase, software applications, standards, and work practices that are used by member agencies. This has greatly improved overall coordination and collaboration among the cities and public utility companies involving construction planning and operations.

3.1.10 Improved Communications, Coordination, and Effectiveness

ROADIS is operated jointly by ROADIC, the road administrators in the national government, the 12 designated cities, and the public utility companies operating within these municipalities. Because the program is federally mandated, cities and public utility companies are required to comply with the ROADIC standards and processes.

This mandate enabled advanced planning, improved coordination, and made construction and protection of infrastructure with the right-of-way more efficient. The result has been minimized disruptions and increased accident prevention, as well as reduced overall planning and construction costs.

The concept of using a common infrastructure database for all members has been discussed extensively throughout the United States over the years. The ROADIC program serves as an example of how such a program could be accomplished in the United States.

3.2 Role of Portals for Disaster Management

Efforts within the United States aimed at improving collaboration among infrastructure stakeholders may be found in one of two forms. The first is a top-down, *closed-system* approach. In this approach, an organization, typically an agency of the federal government, establishes a program for infrastructure information gathering and sharing.

As with the ROADIS, such efforts, discussed later in this chapter, are typically closed and not easily accessed unless a member of the group specifies a requirement or if other, very specific, sets of conditions are satisfied. Closed systems, by nature, often fail to disseminate data to all needed partners.

Conversely, *open systems*—those by which data are made freely available to any potentially interested party—are able to reach broad audiences. Open systems for data sharing are often representative of bottom-up or grassroots approaches. However, such systems typically lack the significant financial support needed to assemble, organize, and disseminate large, accurate, and complete data sets.

Such systems are not compatible with many types of critical infrastructure data due to their open nature and are thus not easily implemented for sharing data without significant "scrubbing" of proprietary or sensitive information. Thus, the ideal system for data collaboration resides in a blend of these two approaches.

For many years, it was thought that geospatial data portals were especially useful for managing emergency responses related to disaster and critical infrastructure failure, and that they could meet the blend of open and closed systems required. Initial thinking by federal and state governments focused on extending geospatial clearinghouses to offer common data sets such as roadways, hydrological features, and political/administrative boundaries.

This goal was intended to be accomplished through the creation and use of password-protected subsections, subnets, or secondary sites. While utilitarian, this approach quickly encountered several challenges beyond the obvious issues of data dissemination and securing password access for thousands of emergency managers and responders.

Events such as Hurricane Katrina redefined and substantially expanded the notion of critical infrastructure data and information required for disaster response. For example, the state veterinarian for Mississippi and the Mississippi Department of Health were in critical need of data depicting the location of henhouses and their surrounding topography after Katrina obliterated electric utility service in 2005.

Their concerns stemmed from the sudden heat-related death of millions of chickens due to a loss of cooling in henhouses and the potential for excess runoff and effluent into nearby drainages and waterways, thereby creating a secondary public health disaster. While limited data were available to support such inquiries, their underlying currency and spatial accuracy were marginal at best for addressing this concern.

Such problems highlighted the rising importance of geospatial technologies to emergency managers, while underscoring the potential difficulties associated with capturing and managing a tremendous diversity of geospatial data. Not only did most emergency management organizations lack the capacity to develop and implement the required data, but also they lacked the financial capability to develop them. In response to this chasm, emergency management organizations began to look with increasing frequency for geospatial support and funding from federal government agencies such as the Department of Homeland Security (DHS) and Federal Emergency Management Agency (FEMA).

Beyond the lack of available data, the portal model proved fallible in other ways. The digital size of data increases in direct proportion to its geographic footprint. Further, the same holds true with the potential number of contributing sources. For example, a tornado affecting a small municipality will likely only require electric utility data for a few thousand customers from one or two providers.

Conversely, Katrina affected millions of customers in Mississippi across a geographic region encompassing 60 counties and 33 service providers. Aggregating such potentially large and disparate data sets within the electric utility industry to indicate outages was a monumental task, and the resulting data set was large and unwieldy enough to make electronic transfer impractical.

Lack of Internet infrastructure is consistently a major limitation that prohibits the use of portals and distributed enterprise model geodatabases. High-resolution imagery covering a small municipality may result in file sizes measured in tens or hundreds of megabytes, used by dozens of professionals. A large incident may cover multiple county-sized areas and result in file sizes measured in tens or hundreds of gigabytes and used by hundreds of professionals. Most geospatial portals are not designed to support such massive time-sensitive demands, and thus their use is impractical.

Consistent connectivity is a related theme. Disasters are such because they often destroy electric utility and communication systems. During Katrina, the most reliable means of moving data, both from the field to operation centers and vice versa, was best accomplished using portable drives, which were hand-carried thrice daily by couriers. Even this process was fallible, as once a drive was connected to a secured network or computer, such as those used by the military, it could not be returned.

Digital video discs (DVDs) were an acceptable alternative when the volume of data was less than a disc's capacity, but moving larger data sets proved recalcitrant. The federal government's prime contractor for mapping found itself delivering sets of 73 DVDs to provide complete databases with imagery. The time that was required to copy and deliver 73 DVDs exceeded the refresh time for new data arriving from sensors in space, in the air, and in the field.

The use of portals also presents a challenge from the data contributor perspective. Open availability or exposure of many critical data sets presents an increased risk for terrorism and acts of sabotage. Further, data providers from utility companies and other infrastructure stakeholder groups rightfully have concerns about the loss of control over proprietary information when shared through a portal or third-party systems. Similar concerns also exist with respect to liability issues.

To help alleviate the above challenges and concerns, DHS, in partnership with the National Geospatial-Intelligence Agency (NGA) and U.S. Geological Survey (USGS), implemented the Homeland Security Infrastructure Protection (HSIP) data set program under the auspices of Homeland Security Presidential Directives (HSPDs) 5 and 8. The HSIP program provides a neutral party (government)–based solution, through which infrastructure stakeholders may contribute data in a secure environment.

These data are combined with federal government licensed data sets and products produced by the private sector to create a massive limited-distribution data set for use by federal emergency managers. Access rights to these data may be extended to state and local emergency managers only through a presidential declaration of disaster or special-case, use-specific circumstances. By statute, failure by users to maintain security for HSIP resources could result in criminal prosecution.

Initially, HSIP was only available for access via a large collection of DVDs prestaged with federal response and management assets and select state-level emergency managers. New data were added to the collection or updated on a quarterly basis at best. This approach, effectively, kept HSIP data out of the day-to-day reach of emergency responders.

These conditions were not improved even when HSIP was migrated to a more formal portal structure. This prevented, and in fact violated, a fundamental tenant of emergency response: train in the same way that you would respond. Thus, the lack of availability for training and preparedness activities limited the effectiveness and return on investment of HSIP data layers and services.

This was partially resolved when HSIP data were separated into HSIP Gold data sets (those data bound by strict confidentiality and licensure agreements) and HSIP Freedom data sets (those data contributed through public sources or open data licensing agreements). However, this replaced one problem—lack of familiarity with data structures—with another, admittedly less contentious problem: the need to synchronize the two data sets during a disaster.

This is not to say that the use of portals in clearinghouses by emergency management is a futile effort. Rather, effective implementation must be accompanied by a keen understanding of both capabilities and limitations. Multi-organization agreements, which advocate data sharing and collaboration with respect to the development of portals, will greatly increase their intended effectiveness.

3.3 NOAA's Digital Coast and Other Portals

3.3.1 Ocean and Coastal Mapping Integration Act

In early 2009, Congress recognized the need for accurate coastal maps and the use of geospatial data and technology to protect the economic, social, and ecological interests of coastal communities. The Ocean and Coastal Mapping Integration Act was introduced in the Senate on January 8, 2009, by Senator Daniel Inouye of Hawaii as S. 174. A similarly named bill was introduced by Representative Madeleine Bordallo of Guam on January 9, 2009, as H.R. 365. The Ocean and Coastal Mapping Integration Act would establish a coordinated and comprehensive federal ocean and coastal mapping program.

The legislation directed the president to create a program to coordinate comprehensive federal ocean and coastal mapping for the Great Lakes and coastal state waters, the territorial sea, the exclusive economic zone, and the U.S. continental shelf. The intent was to establish research and mapping priorities for conservation and management of marine resources and habitats, support ongoing and needed research, and expand ocean and coastal sciences.

The bill also directed the administrator of the National Oceanic and Atmospheric Administration (NOAA) to convene a committee on ocean and coastal mapping to implement the program and coordinate federal oceanic and coastal mapping and surveying activities with other federal efforts, including the Digital Coast, Geospatial One-Stop, and the Federal Geographic Data Committee (FGDC). Further, the bill directed the administrator to include international mapping activities, coastal states, user groups, and nongovernmental entities in these coordination efforts. It also directed the administrator to develop a plan for an integrated ocean and coastal mapping initiative within NOAA, as well as establishing joint oceanic and coastal mapping centers of excellence in institutions of higher education to conduct specified research activities.

3.3.2 NOAA's Digital Coast

NOAA's Digital Coast provides coastal communities and the public with land- and marine-based geospatial data to create detailed coastal maps of the United States. The Digital Coast initiative and website provide a web-based mapping tool that allows users the ability to create maps, generate various coastal development scenarios, and assist decision makers with coastal planning and management. Beyond the data and applications, the site also provides training and examples to support analysis and decision making regarding coastal resources and natural events.

The website was launched by the NOAA Coastal Services Center in 2008. The Digital Coast initiative provides the information needed by those who want to conserve and protect coastal communities and natural resources. Geospatial data provided by Digital Coast includes inventory of imagery, land cover, elevation, benthic habitat, hydrology, and marine boundary data sets.

One of the first projects to be considered under the Digital Coast initiative was the establishment of the Louisiana–Mississippi Digital Coast Initiative. NOAA awarded a grant to the University of New Orleans Pontchartrain Institute for Environmental Sciences to develop a repository of digital information for the Pearl River Basin area of coastal Louisiana and Mississippi.

The project established a web-based digital repository of coastal information consisting of various digital data and map products, and remote sensing images. The tool was made available to various public and private agencies to support a wide range of coastal analysis activities, such as coastal restoration, land usage, and population changes.

The NOAA Coastal Services Center and its Digital Coast partners and contributors have developed a number of coastal-related visualization and analytical tools and associated data sets. The following provides a summary of example tools and data at the time of this publication.

3.3.3 BASINS Climate Assessment Tool by the Environmental Protection Agency

The BASINS Climate Assessment Tool (CAT) is an integrated environmental analysis application that incorporates the use of GIS technology, national watershed data, and watershed modeling tools, including the Hydrologic Simulation Program FORTRAN (HSPF) model, into a single solution. It provides users the ability to assess the combined effects of climate and land use change, and to guide the development of effective management responses to ongoing climate challenges.

3.3.4 CanVis Tool by NOAA

The CanVis tool is a web-based application that allows users the ability to visualize potential impacts of coastal development or climate change. This allows both public and private organizations involved with conservation activities to gain a greater understanding of the impacts of their proposed actions.

3.3.5 Coastal County Snapshots Tool by NOAA

Coastal County Snapshots allows users to select and organize data by coastal county and area of interest to gain an understanding of a county's flood exposure or wetland benefits. This information assists both public and private organizations in prioritizing conservation areas to abate the impacts of floods and maintain the vital ecosystem services that natural resources provide.

3.3.6 Coastal Resilience Tool by the Natural Conservancy

The Coastal Resilience tool provides users the ability to view changes in sea levels and storm surge scenarios to identify economic, social, and ecological impacts for specific geographies with unique coastal issues. It provides decision makers with the ability to develop economic and ecological assessments and potential solutions. It incorporates map service capabilities from the Sea Level Rise Viewer as well as other coastal data.

3.3.7 Coastal Vulnerability Maps and Study by EPA

The Coastal Vulnerability Maps and Study by the Environmental Protection Agency (EPA) resulted in the creation of elevation map data of coastal areas from Massachusetts to Florida, and Texas. This initiative includes a sea level rise planning study that integrated data involving land use, zoning, and anticipated development to determine the potential for shore protection and prevention of inland wetland migration.

3.3.8 Habitat Priority Planner Tool by NOAA

The Habitat Priority Planner is a GIS-based tool for identifying and prioritizing coastal areas for land use planning, conservation, and restoration. The tool can incorporate other data sets, such as climate change data, to assess possible impacts on habitat areas for conservation based on those impacts.

3.3.9 Integrated Climate and Land Use Scenarios GIS Tool by EPA

The Integrated Climate and Land Use Scenarios tool will provide users with the ability to develop future scenarios for land use based on potential population growth projections, greenhouse gas emissions, and socioeconomic changes. EPA was circulating the tool for external peer review at the time of this publication.

3.3.10 Land Cover Atlas by NOAA

NOAA's land cover atlas is a web-based tool for reviewing county-level land cover change over time. Users can identify emerging trends, predicting and assessing cumulative impacts of such challenges as rises in sea level. Condition data can be used to assess resilience and highlight areas where these impacts may need to be addressed. The tool can assist with conservation or land use planning efforts and can improve decision making and communication at the local and regional levels.

3.3.11 National Atlas by U.S. Fish and Wildlife Service

The national atlas produced by the U.S. Fish and Wildlife Service (USFWS) is a mapping tool that provides users the ability to analyze issues related to climate change, carbon levels in the environment, ecosystem changes, and biodiversity loss. These data support a variety of both public and private organizations, including federal, state, tribal, and local government decision makers and other users with an interest in land cover dynamics, conservation, and continental-scale patterns of changing environment.

3.3.12 Open-Source Nonpoint Source Pollution and Erosion Comparison Tool by NOAA

The Open-Source Nonpoint Source Pollution and Erosion Comparison Tool is a GIS-based tool that assists users in identifying land areas that generate high sediment and nonpoint source pollutant loads. The tool includes scenario analysis capabilities to allow users to test the relative effectiveness of different land management options for sediment and pollutant yields. The impacts of changing climatic regimes can also be tested against potential changes to soils, land cover, precipitation, and other relevant characteristics.

3.3.13 Sea Level Rise and Coastal Flooding Impact Viewer by NOAA

The Sea Level Rise and Coastal Flooding Impact Viewer provides a national viewer for allowing communities to understand their potential exposure to inundation from coastal flooding due to storms or a sea level rise. The viewer uses a digital elevation model (DEM) as a base for mapping inundation and storm surge modeling and mapping by the National Hurricane Center. The DEMs can be downloaded for use by local communities to address their planning and response needs.

3.3.14 Spatial Trends in Coastal Socioeconomics Quick Report Tool by NOAA

This reporting tool allows users to gain insights into socioeconomic trends in the developing coastal regions of the United States. This can be used to inform decisions on conservation priorities as a result of the impacts of climate change. The tool uses a map-based interface to determine demographic and economic characteristics for a variety of coastal management jurisdictions. Users can identify key socioeconomic data sets, and create customized reports of demographic and economic data for specific jurisdictions.

3.3.15 State of the Coast Website by NOAA

The State of the Coast website provides statistics through interactive visualizations that highlight known facts about coastal communities, coastal economies, and coastal ecosystems, and how climate change might affect the coast within a particular area. The website includes case studies that consider the interrelationships among the four states of the coast themes: coastal communities, ecosystems, economy, and climate.

3.3.16 The Sea Level Affecting Marshes Model View Tool by USFWS

The Sea Level Affecting Marshes Model (SLAMM) program is designed to help people gain an understanding of the potential impacts of climate change on sea levels. This web-based tool allows the public to view simulations of sea level rise, and displays map comparisons of the same coastal area, at different sea levels. Users can visually see the anticipated modeling of sea level rise and associated impacts.

3.3.17 U.S. Interagency Elevation Inventory by NOAA

The U.S. Interagency Elevation Inventory provides a listing of known topographic and bathymetric data for the United States and its territories. These elevation data are used to develop coastal elevation models that are critical components when analyzing sea level rise, coastal flooding, and erosion.

3.3.18 Digital Coast Partnership

The Digital Coast Partnership represents a diverse group of organizations that actively participate in a variety of collaborative initiatives with NOAA. Their involvement in the creation of information resources contributes to the NOAA tools and services being used to address challenges such as communities and infrastructure at risk to coastal hazards, and strengthen coastal resilience. The partnership also reduces duplication of effort by these organizations in developing delivery systems of this information, as well as the collection and management of coastal-related data. The following include the Digital Coast Partnership and links to their respective websites:

- American Planning Association (http://www.planning.org/)
- Association of State Floodplain Managers (http://www.floods.org/)
- Coastal States Organization (http://www.coastalstates.org/)
- Digital Coast (http://www.csc.noaa.gov/digitalcoast/)
- National Association of Counties (http://www.naco.org/)
- National Estuarine Research Reserve Association (http://nerra.org/)
- National States Geographic Information Council (http://www.nsgic. org/)
- NOAA Coastal Services Center (http://www.csc.noaa.gov/)
- The Nature Conservancy (http://www.nature.org/)
- Urban Land Institute (http://www.uli.org/)

3.4 Data Sharing and the Evolution of the Geospatial Platform

The Digital Coast program represented an important stage in the evolution not only of data aggregation and sharing, but also of public access to those data. Other examples of the portal approach include the cooperative efforts undertaken by FEMA. In 2006, FEMA collaborated with the U.S. Army Corps of Engineers to obtain terrain information developed by the Corps. FEMA inventoried more than 600 elevation data sets and more than 580 imagery data sets and catalogued these data on the FEMA Mapping Information Platform (FGDC, 2006, p. iii).

While these efforts were underway, a much larger and longer-term initiative was building within the federal government. In 1990, the Office of Management and Budget (OMB) established the FGDC as "an interagency committee that promotes the coordinated development, use, sharing and dissemination of geospatial data on a national basis" (http://www.fgdc.gov/).

Executive Order 12906, "Coordinating Geographic Data Acquisition and Access: The National Spatial Data Infrastructure," was signed by President William J. Clinton on April 11, 1994. This executive order launched the initiative to create the National Spatial Data Infrastructure (NSDI), an initiative "to support public and private sector applications of geospatial data in such areas as transportation, community development, agriculture, emergency response, environmental management and information technology."

Rechartered in 2002 by OMB, the interagency FGDC was tasked with coordination of the NSDI. This executive order, and the amending Executive Order 13286 issued by President George W. Bush on March 5, 2003, which included the DHS by specific reference, was realized through the NSDI Clearinghouse.

> The NSDI Clearinghouse Network, sponsored by the FGDC, is a distributed system of agency servers located on the Internet that contain field-level descriptions of available and planned digital spatial data, applications, and services. This descriptive information, known as metadata, is collected in a standard format to facilitate query and consistent presentation across multiple participating sites. Clearinghouse uses standards-based Web technology for the publication and discovery of available geospatial resources through the Geospatial Platform portal. The fundamental goal of Clearinghouse is to provide access to digital spatial data and related online services for data access, visualization, or order. (https://www.fgdc.gov/dataandservices/clearinghouse_qanda)

The Geospatial One-Stop (GOS) portal was designated as the official means to access the metadata resources managed in the NSDI Clearinghouse Network. GOS

> saw an increase of more than 30 percent of the accessible metadata records during FY 2006. The GOS Partnership Marketplace, which allows organizations to publish their intent in collecting geospatial data, grew to include approximately 3,000 planned data acquisition records by the end of FY 2006. For more information, see pages 14 and 15. (FGDC, 2006, p. ii)

The clearinghouse continued to grow and GOS continued to provide a method of access for several years. However, the rapid growth of cloud computing and related data access technologies (for example, online data services) began to raise questions regarding the long-term viability of this approach.

In 2008, the National Geospatial Advisory Committee (NGAC) was created within the framework of the Federal Advisory Committee Act (Public Law 92-463). This act defines "a structured process for creating, operating, and terminating Federal advisory committees that provide advice to the Executive Branch of government" (http://www.gsa.gov/portal/category/101111). The mandate of the NGAC was to provide advice and recommendations to the FGDC.

As noted by Folger (2009), one of the first recommendations of the NGAC was contained in its report *The Changing Geospatial Landscape* (NGAC, 2009). In this report, the NGAC noted that "as geospatial data production has shifted from the federal government to the private sector and state and local governments, new partnerships for data sharing and coordination are needed." The NGAC specifically called for better methods of sharing data:

The relative shifts in data production from the federal government to the private sector and state and local government call for new forms of partnership. Furthermore, the hodgepodge of existing data sharing agreements are stifling productivity and are a serious impediment to use even in times of emergency. There is an urgent need to reexamine the relationships between data providers and users to establish a fair and equitable geospatial data marketplace that serves the full range of applications. When the federal government was the primary data provider, regulations required data to be placed in the public domain. This policy jump-started a new marketplace and led to the adoption of GIS capabilities across public and commercial sectors. However, these arrangements are very different when data assets are controlled by private companies or local governments. (NGAC, 2009, p. 12)

The incorporation of data provided by nonfederal sources into a federal data repository created several challenges for the NSDI and the FGDC. Combined with the rapid changes in technologies mentioned previously, these challenges prompted a thoroughgoing review of the systems in place. This review resulted in the recommendation to overhaul the approach to data sharing and create an innovative environment for comprehensive access to geospatial data—the Geospatial Platform.

The Data.gov website was designed to provide greater transparency in government and broad access to data. As the website states, "Open data is fuel for innovators. It has the potential to generate more than $3 trillion a year in additional value in sectors including finance, consumer products, health, energy and education, according to a recent study." At the writing of this text, more than 130,000 federal data sets were available to the public.

Initially, the Geospatial Platform was conceived as a counterpart to Data.gov, albeit a counterpart focused on data with geospatial content (Figure 3.1). As quickly became apparent, relatively few federal data sets

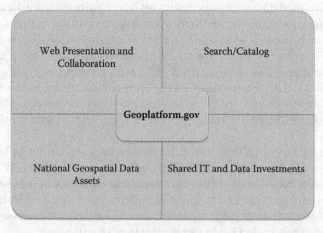

FIGURE 3.1
Geoplatform.gov. (From http://www.geoplatform.gov/. With permission.)

are without geospatial content. At the writing of this text, more than 93,000 federal data sets are already registered and available on the much younger Geospatial Platform (http://www.geoplatform.gov/) via web services. The second issue—the availability of nonfederal data sets, including crowd-sourced data, through the same site—remains a work in progress.

The Geospatial Platform, which has come to be known more simply as the Geoplatform, is a major initiative of the FGDC. The Geoplatform features a shared technology environment that enables the publication and organization of geospatial data provided by government agencies and their trusted partners. This effort is a component of the federal government's Information Technology (IT) Shared Services Initiative and is designed to help agencies more effectively produce and share their geospatial data, services, and applications across the government and with their external partners.

The Geoplatform initiative is organized around four major components:

- Facilitating collaboration and web presentation of geospatial content
- Discovery of spatial data and tools
- Supporting the establishment of and reporting on national geospatial data assets (NGDAs)
- Supporting shared information technology and data investments

Plans for the future development of the Geoplatform include

- Expanding the use of cloud computing
- Implementing a data as a service (DaaS) offering within a Geoplatform marketplace
- Expansion of Geoplatform.gov portfolio management support and reporting capabilities

Of great significance for critical infrastructure protection, the Geoplatform has been designed to be responsive to Executive Order 13286 and its specific reference to the Department of Homeland Security. Working closely with the DHS Geospatial Management Office (GMO), the Geospatial Platform has released the new online version of the Homeland Security (HLS) Geospatial Concept of Operations (GeoCONOPS) (http://www.geoplatform.gov/geoconops-home) (Figure 3.2).

> This Homeland Security (HLS) Geospatial Concept of Operations (GeoCONOPS) is a strategic roadmap to understand, and improve, the coordination of geospatial activities across the entire spectrum of the Nation: from federal, to state, and local governments, to private sector and community organizations, academia, the research and development industry and citizens in support of Homeland Security and Homeland Defense (HD). (https://www.geoplatform.gov/node/575)

Geospatial CONOPS Community Model

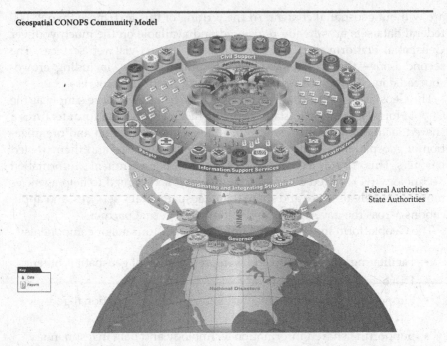

Federal Authorities
State Authorities

FIGURE 3.2
(See color insert.) Homeland Security (HLS) Geospatial Concept of Operations (GeoCONOPS).
(From Homeland Security Geospatial Concept of Operations [GeoCONOPS]. With permission.)

In many respects, the Geoplatform is the penultimate step in the development of national geospatial data sharing. The addition of (or at least access to) nonfederal data represents the final step in achieving this goal.

References

Bush, G.W. (2003). Executive Order 13286 of February 28, 2003: Amendment of executive orders, and other actions, in connection with the transfer of certain functions to the secretary of Homeland Security. *Federal Register*, 68(43).

CADD/GIS Technology Center for Facilities, Infrastructure and Environment. (1996). *Tri-Service Spatial Data Standards ARC/INFO Technical Implementation Guide.*

Clinton, W.J. (1994). Executive Order 12906 of April 11, 1994: Coordinating geographic data acquisition and access: The national spatial data infrastructure. *Federal Register*, 59(71).

Federal Geographic Data Committee. (2006). *2006 Annual Report.* http://www.fgdc.gov/library/whitepapers-reports/annual%20reports/2006-report/index_html.

Folger, P. (2009). *Geospatial Information and Geographic Information Systems (GIS): Current Issues and Future Challenges*. R40625. Congressional Research Service, June 8.

National Geospatial Advisory Committee. (2009). The changing geospatial landscape. January.

4

Industry's Application of GIS to CIP

4.1 Private Ownership of Critical Infrastructure

The National Infrastructure Protection Plan (NIPP) "outlines how government and private sector participants in the critical infrastructure community work together to manage risks and achieve security and resilience outcomes" (DHS, 2014). The NIPP was created in response to directives issued by Presidents George Bush (2003) and Barack Obama (2013). One of the more significant requirements of the NIPP was the definition of sector-specific plans for critical infrastructure protection.

The U.S. Government Accountability Office (GAO) is an independent, nonpartisan agency that works for Congress. The agency's mission statement is "to support the Congress in meeting its constitutional responsibilities and to help improve the performance and ensure the accountability of the federal government for the benefit of the American people. We provide Congress with timely information that is objective, fact-based, nonpartisan, nonideological, fair, and balanced."

One issue examined by the GAO on several occasions in recent years is the degree to which Department of Homeland Security (DHS) efforts satisfied the requirements of the NIPP. For example, in 2007, the GAO issued a report entitled *Critical Infrastructure: Sector Plans Complete and Sector Councils Evolving* (GAO, 2007). In that report, the GAO noted:

> As Hurricane Katrina so forcefully demonstrated, the nation's critical infrastructures—both physical and cyber—have been vulnerable to a wide variety of threats. Because about 85 percent of the nation's critical infrastructure is privately owned, it is vital that public and private stakeholders work together to protect these assets. (GAO, 2007, p. i)

The GAO also noted:

> Representatives of the government and sector coordinating councils had differing views regarding the value of sector-specific plans and DHS's review of those plans. While 10 of the 32 council representatives GAO interviewed reported that they saw the plans as being useful for

their sectors, representatives of eight councils disagreed because they believed the plans either *did not represent a partnership among the necessary key stakeholders, especially the private sector* or were not valuable because the sector had already progressed beyond the plan. (GAO, 2007, p. i, emphasis added)

In a subsequent report, the GAO wrote:

The Conference Report accompanying the Department of Homeland Security Appropriations Act, 2005, directed DHS to complete an analysis on whether the department should require private sector entities to provide DHS with existing information about their security measures and vulnerabilities to improve the department's ability to evaluate critical infrastructure protection nationwide.... The analysis was to include all critical infrastructure, including chemical plants; the costs to the private sector for implementing such a requirement; the benefits of obtaining the information; and costs to DHS's Information Analysis and Infrastructure Protection (IAIP) (presently the Office of Infrastructure Protection (IP)) to implement this requirement. (GAO, 2009, pp. 2–3)

The 2009 GAO report also noted that DHS "was involved in developing a public-private partnership structure" (p. 4) to record information about critical infrastructure and key resources (CIKR). Because

the private sector owns approximately 85 percent of the nation's CIKR— banking and financial institutions, telecommunications networks, and energy production and transmission facilities, among others—it is vital that the public and private sectors work together to protect these assets. (GAO, 2009, pp. 1–2)

The Strategic Foresight Initiative on the Critical Infrastructure (SFI) facilitated by the Federal Emergency Management Agency (FEMA) issued "Long-Term Trends and Drivers and Their Implications for Emergency Management." In that report, SFI noted, "The private sector owns the vast majority of the Nation's critical infrastructure and key resources—roughly 85 percent" (FEMA, 2010, p. 2).

In summary, the consensus of multiple government agencies, as expressed over a period of several years, is that the vast majority of the nation's critical infrastructure is owned, operated, and maintained by the private sector. The private sector is, of course, well aware of this responsibility, as the nation's telecommunications companies, power companies, railroad companies, pipeline companies, private road builders, and numerous other entities can attest.

Indeed, the private sector has communicated its concerns about this responsibility, and Austin (2005, 2010, 2012), Gomez (2008), and Austin et al. (2010), among others, have presented these concerns on numerous occasions. Indeed, this was one of the key concerns that prompted the creation of the

Automated Mapping/Facilities Management (AM/FM) conferences, which evolved into the Geospatial Information and Technology Association (GITA).

4.2 Genesis of AM/FM for Infrastructure Design and Protection

In the late 1960s, a group of engineers and managers working at the Public Service Company of Colorado were among the first to develop a record and information management system that combined automated, computer-assisted mapping and a database of information about the company's facilities. This AM/FM system was designed to assist employees in designing, building, operating, maintaining, and managing the company's investments in infrastructure. It was revolutionary for its time and provided significant advantages in efficiency and cost-effectiveness for the company.

People saw the technology's potential, and plans were made to share information about this approach to information management. Founded by Henry A. Emory and sponsored by the Kellogg Corporation consulting firm, the first AM/FM conference was held in 1978 in Keystone, Colorado, attracting 32 attendees (Figure 4.1).

When the fourth Keystone conference was held in 1981, more than 200 attendees participated. After the conclusion of the fifth Keystone conference in 1982, a formal, not-for-profit organization was chartered to serve this growing industry with an educational forum to exchange ideas and keep up with rapidly advancing technologies. The association was named AM/FM International, reflecting a revolutionary new technology that was sweeping the utility industry (Figure 4.2).

AM/FM International provided vision and leadership in educating individuals interested in implementing AM/FM and geospatial technology. The conferences attracted considerable participation from Europe; a European division was formed in 1984, and the first European division conference was held in Montreux, Switzerland, in 1985 (Figure 4.3).

By 1987, membership in the North American division of AM/FM International reached 1,305. Plans were made to transition the management team from a combination of contractors and volunteers to a full-time staff. During the 1988 conference in Snowmass, Colorado, Robert Samborski was hired as the association's executive director.

Through 1988, the conferences had focused primarily on user and vendor presentations about the technology. Although a few vendors had technology on display in their hotel rooms, no exhibit halls had been organized or supported. That changed in 1989, when the conference was moved to New Orleans and drew 1,329 participants (Figure 4.4).

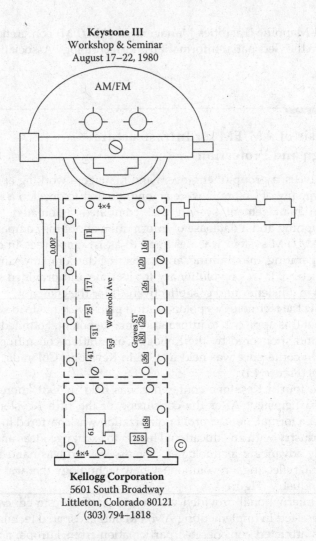

FIGURE 4.1
Keystone Conference III program. (From GECCo and GITA PowerPoint slides. With permission.)

FIGURE 4.2
AM/FM logo. (From GECCo and GITA PowerPoint slides. With permission.)

FIGURE 4.3
Montreux conference. (From GECCo and GITA PowerPoint slides. With permission.)

With the establishment of an exhibition to accompany formal presentations, it became possible for participants to gain firsthand understanding of the profound differences in technologies discussed in Chapter 2. Significantly, this understanding could be gained with a minimal investment in conference attendance, as opposed to the far more expensive processes of issuing Requests for Information (RFIs), Requests for Quotations (RFQs), or Requests for Proposals (RFPs), which were the only viable alternative in a precommercial Internet world.

During the 1990s, AM/FM International continued to grow: internationally, with new divisions and conferences in Japan and Australia/New Zealand, and domestically, with the addition of numerous regional chapters in the United States and Canada. It also grew systemically, with the addition of conferences devoted to executive management and forums dedicated to newer, related technologies, such as supervisory control and data acquisition (SCADA) systems, and organizationally, with the establishment of a permanent secretariat. At the same time, multiple revolutions in geospatial technology were underway, known by some as the second great wave of geographic information systems (GIS) innovation.

In 1998, AM/FM International changed its name to the Geospatial Information and Technology Association (GITA) to reflect the association's new and expanded focus and membership (Figure 4.5). GITA's mission is to provide excellence in education and information exchange on the use and benefits of geospatial information and technology in telecommunications, infrastructure, and utility applications worldwide. GITA's membership

FIGURE 4.4
New Orleans conference. (From GECCo and GITA PowerPoint slides. With permission.)

FIGURE 4.5
GITA logo. (From GECCo and GITA PowerPoint slides. With permission.)

and services were targeted at infrastructure-based organizations that benefit from the application of geospatial information technologies, including electric utilities, gas utilities, telecommunications companies, water and wastewater utilities, public works agencies, local government, and oil and gas pipelines.

By 2000, the organization's membership had grown to include more than 2,200 individuals, 140 user affiliates, and 150 vendor companies. The 2000

FIGURE 4.6
GITA's 2000 exhibition hall. (From GECCo and GITA PowerPoint slides. With permission.)

annual conference attracted more than 3,800 industry professionals who participated in almost 100 educational sessions. A 100,000-ft^2 exhibition hall displayed hardware, software, and services provided by more than 140 vendor companies (Figure 4.6).

As geospatial technology matured, the original AM/FM and GITA focus on evaluation and selection of hardware and software systems became less relevant. Taking its place were concerns about the integration of geospatial technologies with other corporate information systems. In particular, the relevance of GIS for infrastructure management and protection became quite clear.

In 2005, GITA made a presentation to the Homeland Infrastructure Foundation-Level Database (HIFLD) Working Group meeting held at the Federal Communications Commission (FCC) headquarters in Washington, D.C. In that presentation, GITA described the Community Framework for Critical Infrastructure Protection: The GECCo Initiative, which was an outcome from interaction between the Japan division and the North American division described in Section 3.1 (Austin, 2005). This presentation engendered considerable interest on the part of federal government agency employees who were in attendance.

In 2008, GITA organized a Geospatial Dimensions of Emergency Response Symposium in Seattle (Figure 4.7). In 2009, GITA organized a second symposium in Tampa, Florida. This symposium was designed to highlight the following issues:

FIGURE 4.7
Geospatial dimensions of Emergency Response Symposium. (From GECCo and GITA PowerPoint slides. With permission.)

- Applications of geospatial technologies to emergency and disaster response
- Closing the gap between the geospatial and emergency response communities
- Better understanding of the roles and responsibilities of each community
- Encouraging and maintaining an ongoing dialogue to provide for better efficiency in times of disaster and emergency

Working with numerous federal agencies with responsibility for critical infrastructure protection, including the National Geospatial Advisory Committee (NGAC), the Federal Geographic Data Committee (FGDC), and the Bureau of the Census, GITA has provided a conduit for information flows between the private sector and governments. Shortly after the 2009 Emergency Response Symposium, GITA briefed the HIFLD Working Group with an update (Austin, 2010) on the Geospatially Enabling Community Collaboration (GECCo) initiative.

4.3 Adaptation of ROADIC to U.S. Circumstances by GITA

As discussed in Section 3.1, in November 2003, a research team sponsored by GITA met with executives of the ROADIC initiative, along with several key ROADIC members, to learn how the Japanese national government was leveraging GIS technology to support the management and protection of its infrastructure. The ROADIC program was set up as a national project to manage and protect the public utilities within the rights-of-way (ROWs) of the 12 major urban centers throughout Japan. The center coordinates with local government agencies and public utility companies, including electric, gas, water, sewer, trains, subways, and communications.

The concepts identified during the study mission fascinated the GITA research team because infrastructure involves geographically distributed networks of physical assets, and the organizations involved have had to cooperate from both a local and a national perspective to build, maintain, and manage infrastructure to share, analyze, and display vital infrastructure information. In particular, the GITA research team analyzed the structure of the ROADIC program to identify key components that could be adapted to U.S. circumstances. The team identified seven focus areas:

1. **Coordination activities.** There is a formal structure for coordinating activities involving research of use of road space, road and emergency planning, administration and protection of infrastructure, and road work coordination between the federal and local government agencies and public utility companies.

2. **Public safety.** The ROADIC program was created to ensure public safety. Due to a series of gas explosions, ROADIC, and subsequently ROADIS, was created to comprehensively manage and integrate the disparate data about roads and utilities within the ROW using GIS technology so that road administrators and public utility companies could better safeguard the public during a disaster, and subsequent response and recovery activities.

3. **Critical infrastructure protection.** ROADIC provides a key aspect of home island security in the area of critical infrastructure protection. It is enabling government agencies and public utilities to better coordinate and share vital infrastructure information for disaster planning and recovery activities.

4. **Organizational effectiveness.** Member effectiveness has resulted from the improved overall collaboration and coordination among federal agencies, cities, and public utility companies involving infrastructure management, construction planning and operations, and sharing of common data.

5. **Risk management.** Because ROADIS is a closed system, data and system access is limited to those organizations that are members. Restricting access to data and the system is enforced by a membership agreement with each member. This agreement includes strict system and data security policies.

6. **Shared reduced costs.** Development of the common landbase map, access to member public utility company data, and ROADIS applications are the primary ways the ROADIC program continues to reduce costs among member organizations.

7. **Common standards.** Common standards have been established for the organization of infrastructure data, design and mapping standards, work practices, and software applications that are required of all member organizations.

The founders of the ROADIC program realized that GIS data and technology provided the greatest potential for analyzing the spatial relationship of assets, resources, and people, and were ideally suited to support many of the complex interrelationships involved in the design, management, operation, and protection of their infrastructure. For ROADIC, GIS had become a key tool for supporting infrastructure management and protection because it facilitates collection, retrieval, organization, and integration of information.

In a findings report, the GITA research team wrote that the ROADIC program was an extraordinary GIS-based program that supports the coordination and protection of a large portion of Japan's infrastructure. Furthermore, the report noted that ROADIC's ability to effectively administer and protect infrastructure helped both local and federal agencies make more informed, timely, and cost-effective decisions when planning for and responding to natural or man-made events.

Based on the research team findings, GITA's Board of Directors established the National Geospatial Initiative for Critical Infrastructure Protection in 2004. The initiative included the formation of a CIP Task Force. Their initial work focused on the evaluation of Presidential Directives HSPD 5, HSPD 7, and HSPD 8 associated with the protection of critical infrastructure within the United States.

The task force began holding meetings with representatives from the DHS, the Transportation Security Administration (TSA), the FGDC, and the National Geospatial-Intelligence Agency (NGA). These meetings were designed to share the findings from ROADIC and the ROADIS program and to learn about the various federal agency efforts underway focused on establishing standards and mandates, collecting and organizing critical infrastructure data, and developing various GIS-based software applications and databases.

During the following year, the CIP Task Force generated a number of templates and techniques focused on critical infrastructure protection based on the lessons learned from ROADIC and various federal agency programs focused on those presidential directives. The eight specific documents and templates created by the task force are:

1. Model letter of intent and data-sharing memorandum of agreement
2. Model confidential agreement
3. Example data-sharing and collaboration project purpose and partnering process
4. Example critical infrastructure interdependencies material and exercises
5. Model data-sharing and collaboration project overview presentation material
6. Example emergency management and critical infrastructure protection recommendations

7. Example GIS applications for emergency management and critical infrastructure protection

8. Model guideline considerations for spatial data access and sharing

The development of this material, as well as the ongoing discussions with DHS, culminated in the formation of a GITA pilot concept, which later became known as the Geospatially Enabling Community Collaboration (GECCo). The charge of the GECCo program was to facilitate a series of pilot projects to be held across the United States.

Each pilot was done in cooperation with representatives of the DHS, the FGDC, and other federal, state, and local agencies, as well as the private utility and telecommunications industries. Pilot projects included a combination of infrastructure stakeholders and emergency responders in a local or regional area for the purpose of addressing collaboration and data-sharing issues that inhibit effective critical infrastructure protection and emergency response in times of a natural disaster or man-made event.

4.4 The GECCo Program

4.4.1 Protecting Critical Infrastructure

Homeland security will remain a national priority by our government for the near future. However, beyond the obvious impact of potentially successful terrorist attacks, it is important to remember that the results of natural disasters are just as serious. Hurricanes, earthquakes, tsunamis, floods, and fires occur with unpredictable regularity and significant cost in lives and property. Damage to underground infrastructure by excavators occurs on a daily basis. While most of this accidental damage goes unnoticed on the national level, the aggregate effect on the economy is staggering, and the number of lives lost is tragically unnecessary.

Regardless of the cause of the emergency—terrorism, natural occurrences, or unintentional human error—the methods of responding to, mitigating, and ideally preventing reoccurrences are based on a common approach: the coordinated use of spatial information. This cannot happen without the many mutually dependent agencies and organizations charged with protecting the infrastructure and the vital need to efficiently and effectively share their GIS data. There are obstacles that need to be overcome before this collaboration can occur, however, and that was the primary impetus behind the GECCo program (Figure 4.8).

From a GECCo perspective, a community depends on critical infrastructure for such things as economic security, quality of life, delivery of service, and governance. As we have experienced in numerous disasters, the denial

FIGURE 4.8
GECCo logo. (From GECCo and GITA PowerPoint slides. With permission.)

of one or more of these has a profound negative effect on both the public and private sectors within that community.

As a result, it is important to identify the interconnectivity among critical infrastructure and their supporting systems within a community, to understand not only their vulnerability, but also their ability to withstand and recover from disruptions. The importance of understanding critical infrastructure interactions and vulnerability has taken on a new urgency due to the increase in natural events and terrorist activities. The GECCo Pilot Project Program provided value to the communities and organizations that participated in each of the events because it helped define required information, identified data exchange and collaboration models, and defined technical solutions that can assist communities in meeting the challenges associated with protecting their critical infrastructure.

The availability of information about critical infrastructure affects its vulnerability, as well as the ability of a community to function, meet its citizen's needs, and grow. The disruption of critical infrastructure by either natural or man-made disasters can change the fundamental characteristics of the community, depending on the community's response. The information related to critical infrastructure may be considered an independent critical infrastructure set of data in and of itself because of its importance to the community.

Both content and access must be protected. Limitation of access to data about critical infrastructure and the infrastructure itself must be balanced against the need for access required to protect information technology.

There are important data to be gathered, research to be conducted, and policies and agreements to address to protect and enhance the critical infrastructure of a community. An important debate that is ongoing is the concern over how much information should be readily available among the stakeholders when planning for and responding to an event. This debate centers around the limits that should be placed on access and sharing of information involving critical infrastructure, to reduce the vulnerability to terrorism, and competitive and safety aspects associated with the protection of both public and private infrastructure within the community.

Finally, the cost of protecting critical infrastructure could overwhelm a community if good decisions are not made concerning the allocation of resources and efforts, and if adequate steps are not taken to reduce the risk of disruption, assess the vulnerability, and develop methods for responding to, mitigating, and preventing occurrences. This requires an understanding of the processes that affect or are affected by the critical infrastructure, the dynamic nature of the threat, natural or man-made, and the data and information needed to build robust mitigation, readiness, response, and recovery capabilities to make the community more resilient. Hence, GITA's call to action on behalf of the infrastructure management community was to lead the development of a national initiative for critical infrastructure protection through the GECCo program.

4.4.2 GECCo Overview

As noted earlier, the premise behind the GECCo program was to conduct a series of pilot projects around the country to facilitate dialogue among infrastructure and emergency management stakeholders in a defined geographic area to begin addressing collaboration and data-sharing issues that inhibit effective critical infrastructure protection in times of emergency.

The 10 U.S.-based GECCo pilot locations, in the order of completion, were as follows:

1. City and county of Honolulu, Hawaii
2. City of Denver and the Front Range, Colorado
3. Western New York State, Southern Tier West Regional Planning Agency
4. City of Seattle and King County, Washington
5. Greater Tampa Bay area, Florida
6. Greater Phoenix area, Arizona
7. Greater Dallas–Ft. Worth, Texas
8. Greater Twin Cities area, Minnesota
9. Oakland and San Francisco Bay area, California
10. City of Charlotte and Mecklenburg County, North Carolina

Specifically, the GECCo program served as cooperative forums within each one of these communities to identify and address intra- and interorganizational collaboration and coordination, effective practices and guidelines, information access and exchange, interoperability and enterprise architecture, and data and technology requirements. In addition, the GECCo program was designed to support jurisdictions at all levels of government, the private sector, and nongovernmental organizations in complying with specific elements of the National Incident Management System requirements, thus enabling community stakeholders to more effectively prepare for, prevent, respond to, and recover from domestic incidents, regardless of cause, size, or complexity.

The outcome of each pilot project was designed to enhance existing security and emergency management–related efforts and enable community stakeholders to develop a framework by which public and private organizations can better collaborate to protect critical infrastructure and respond to emergency situations more effectively. Ultimately, the pilot projects resulted in a set of recommendations for using GIS data and technology and other related best practices in communities across the United States for protecting critical infrastructure and supporting emergency management activities.

4.4.3 Summary of GECCo Findings and Recommendations

The unwillingness to share GIS data is by no means universal. However, local governments and private utility companies are oftentimes less willing to participate. In summary, the typical barriers to collaboration and data sharing that were identified during the pilot projects are

- Fear that once others are aware of the existence of data, they may attempt to obtain access through freedom-of-information laws
- Concern over liability issues associated with providing data to external organizations and concern over what will be done with the data and who the data may be shared with
- Efforts required to convert data into a form that can easily be shared with other stakeholders
- Unwillingness for private infrastructure management companies to share potentially competitive information and locations of their critical infrastructure
- Concern over security involving sensitive information and fear of it getting into the wrong hands

The following findings and recommendations reflect the input provided by the more than 750 participants over the course of conducting the 10 GECCos across the United States, as well as detailed evaluations from leading practitioners, infrastructure personnel, and experts from industry, universities, and local, regional, state, tribal, and federal government agencies. These key stakeholders have made significant strides in developing an unprecedented level of cooperation, working collectively to improve local and regional preparedness across their communities.

These public and private stakeholders were also intent on identifying the challenges they needed to address to make their communities resilient to major disasters that put the region at risk. As a result, participants for each GECCo came to the exercise prepared to contribute, share, and learn. In each case, the stakeholders were candid in their observations and thoughtful and innovative in their recommendations on how to deal with the shortfalls they saw during each of the exercises they participated in.

4.4.3.1 Interdependencies in a Major Disaster

4.4.3.1.1 GECCo Findings

Large-scale disaster scenarios were challenging no matter the location, even for those participants who had attended previous disaster events and had an understanding of regional infrastructure interdependencies and emergency management coordination. Nearly all of the GECCo events identified the issue that participants had little to no real idea of what would be the physical impacts on their facilities, operational and business systems, and components—particularly those that were underground, for example, power and communications cables, and water, wastewater, and fuel and natural gas pipelines.

Although each exercise involved a long-term power outage, the extent of the consequences was not apparent to participants beyond some of the obvious high-level interdependencies. Many individuals appeared not to understand how cascading and simultaneous infrastructure failures and physical destruction of critical assets could paralyze portions of a region for days and weeks.

It was interesting to learn that participants generally lacked knowledge or understanding of how infrastructure interdependencies could greatly exacerbate the effects of a major disaster. Participants did not fully appreciate secondary dependencies, such as the need for IT systems, shelters, fuel delivery, and alternate sources of necessary supplies and products. Many participants from the various GECCos did not take into account how localized damage to or destruction of critical infrastructure assets (for example, electric power substations, dams, and bridges essential to regional transportation) could lead to catastrophic consequences and long-term restriction of essential services.

Overall, participants had difficulty envisioning how local or regional impacts could greatly impede response, recovery, and longer-term restoration activities. During each GECCo, government officials and private sector representatives explained how they would manage the disaster and establish priorities based on their respective response and contingency plans.

With few exceptions, there was recognition that these plans could be compromised by an extensive long-term power outage; the absence of most communications capabilities; major transportation constraints from damaged bridges, tunnels, and roads; water disruption; sewer backup; the shutdown of fuel and natural gas distribution; and fires around the region caused by ruptured gas pipelines. Infrastructure owners often acknowledged they would have difficulty locating and transporting needed materials to rebuild their systems and would be competing with other sectors for limited heavy equipment and operators. See, for example, GAO, 2013.

Most participants failed to appreciate the monumental issue of rescuing hundreds or thousands of individuals affected by a disaster, the need to provide shelter for or resettle thousands of others, and dealing with the dead from a major disaster. The issue of how to bring in response and recovery resources from outside a region to help was often raised, as was the need for staging areas for resources and the relocation of large numbers of people.

As for prioritization of service restoration, there was general recognition that there needed to be a systematic way to prioritize based on changing response and restoration needs of critical infrastructures and essential service providers. The question of who establishes restoration priorities and who resolves the conflict over competing priorities was raised during each GECCo event. Participants consistently raised concerns that if emergency operations centers (EOCs) were located within a disaster area, many organizations would not be equipped to establish a backup if the EOC was lost or unavailable.

4.4.3.1.2 GECCo Recommendations

1. Develop a structure process to identify and assess the importance of regional interdependencies.

2. Explore what assessment tools might be available that address interdependencies (for example, Hazus and capabilities of the DHS National Infrastructure Simulation and Analysis Center and other national laboratories and research institutions), with particular focus on those that use GIS that could be used for preparedness and disaster response.

3. Update and improve existing federal, tribal, state, regional, and local preparedness and disaster management plans to address interdependencies in a major disaster scenario.

4. Incorporate interdependencies into vulnerability and emergency response and contingency plans to take into account interdependencies and related restoration needs. This should include mitigation strategies, priorities, and service restoration.

5. Encourage critical infrastructure owners and essential service providers to establish alternative sources for essential products and services—for example, for water systems, alternative sources of drinking water and alternative methods of water distribution.

6. Develop a regional agreement of service restoration priorities for all critical services (for example, electrical, water, oil, and gas), and address the issue of who makes that decision and which organization or organizations can reprioritize service restoration during the course of response and recovery.

7. Examine evacuation and sheltering plans for thoroughness, taking regional interdependencies into account, as well as shelter limitations and vulnerabilities, using a major disaster scenario as a baseline.

8. Conduct workshops and exercises, both sector specific and regional, including field exercises, involving public–private organizations to examine interdependencies at deeper levels to identify gaps and solutions.

4.4.3.2 Geospatial and Information Technology and Telecommunications

4.4.3.2.1 GECCo Findings

In general, exercise participants repeatedly had difficulty dealing with the fact that damage and disruption of telecommunications and critical information technology assets left much of a region without emergency and general communications capabilities and operational systems. Public safety infrastructure in a region is dependent on telecommunications carriers and the 800 MHz network; SCADA and other process control systems in a wide range of infrastructures would also be damaged or disrupted.

While all of the GECCo exercises demonstrated the need for interoperable communications, a reoccurring issue was the impact of the loss of telecommunications and critical IT systems and how these systems could be made more resilient. Some participants pointed to mitigation measures, including building more systems redundancy and developing alternative, mobile, and easily deployable wireless-based communications across a region.

Participants agreed that there was a need for greater knowledge of what was happening throughout a region as the exercise unfolded to enable optimal decision making on response (for example, dispatching personnel and other resources where needed, prioritizing service restoration, and determining evacuation routes and sheltering locations).

In each of the exercise scenarios, government agencies were forced to suspend operations of their computer-based services, along with those of other organizations, cutting off essential IT-associated local government services until essential services (for example, power, water, telecommunications) could be restored, equipment repaired, and systems put back online.

In nearly all GECCos, it was unclear what state and federal agencies had to offer regarding assistance to public and private sector organizations to respond to loss or damage to operational or business systems. It was consistently pointed out that under the National Response Plan (NRP), the National Communications Service in DHS has this role for telecommunications; for critical IT infrastructure, the lead agency is not designated.

Emergency response and business contingency plans generally assumed the existence of communications channels and lack of provisions to take into account the absence of telecommunications, cellular communications, and critical IT infrastructure, or its disruption, damage, or destruction.

IT and telecommunications assets are often co-located in the same building, which increases vulnerability. Although they may have backup systems, including generators and fuel, policies or other restrictions may impede them from sharing these resources during a disaster.

4.4.3.2.2 GECCo Recommendations

1. Create a GIS working group to work on approaches and mechanisms to improve geospatial information sharing and collaboration. Conduct emergency management exercises leveraging GIS technology and involve the emergency operations centers, law enforcement personnel, and other relevant agencies to gain understanding and obtain buy-in for leveraging the use of GIS to support emergency management needs.

2. There is a need to use GIS for situational awareness—knowledge of what is happening throughout the region—as a disaster unfolds to enable optimal decision making on response (for example, dispatching personnel and other resources where needed, prioritizing service restoration, and determining evacuation routes and sheltering locations).

3. Provide education to potential stakeholders, politicians, first responders, and emergency operations staff to help them see the benefits of GIS technology and geospatial data.

4. Develop a risk assessment methodology for telecommunications/critical IT infrastructure resiliency, along with criticality criteria to prioritize telecommunications and IT infrastructure assets. The assessment should include a baseline inventory of government, private sector, and other essential primary telecommunications systems, including those used for emergencies, and include mitigation alternatives to address identified vulnerabilities and alternate communications links if disrupted.

5. Encourage all organizations to include within their contingency plans provisions for backup systems to ensure redundancy to deal with outages of phone, mobile phone, and Internet access, and explore greater use of high-speed Internet voice and data, customer contact, hotline numbers, satellite phones, and text messaging for disaster response.

6. Develop a public–private sector plan for a regional telecommunications/critical IT infrastructure system that ensures interoperability and compatibility among stakeholder communications and IT systems. Incorporate this plan into an updated state NRP Emergency Service Function (ESF) 2.

7. Where appropriate, key stakeholder representatives should share phone numbers, radio frequencies, and other contact alternatives, within sectors and cross-sector with critical customers, service providers, contractors, and others deemed necessary to meet contingency planning requirements for their organization.

8. Identify locations and amounts of necessary emergency equipment, such as power generators, extended-life batteries, and standardized charger connections. Consider how to enlarge emergency fuel supplies for generators and emergency vehicles. Encourage telecommunications companies to explore with co-located companies the sharing of stockpiled resources not allowed by current contracts.

9. Provide access for interested public and private sector organizations to the government emergency networks, Wireless Priority Service (WPS), and Telecommunications Service Priority (TSP) program to enable the sharing of information, tools, and expertise in a regional disaster.

10. Secure means to provide technical expertise for telecommunications and critical IT infrastructure assessment and disaster preparedness/management.

11. Investigate ways to link first responders and local and private sector EOCs to local radio stations to provide to the public notification of outages, threat information, and general information when phone lines, common networks, and email are not available.

12. Link regional EOCs, including utility EOCs, through a regional communications network based on resilient interoperable systems, such as radio, satellite phone, and IT capabilities.

13. Encourage organizations to establish a schedule to ensure routine testing of existing communications systems and incorporate it into in-house organization exercises.

4.4.3.3 Coordination and Cooperation

4.4.3.3.1 GECCo Findings

Most GECCo exercises revealed that key stakeholders had some level of public–private cooperation already developed, holding a major disaster exercise and meeting in a number of forums, as well as holding other exercises and workshops to improve preparedness and response.

Many of the GECCo participants that had disaster management responsibilities within their organization knew each other and occasionally shared information. All participants were very interested in advancing their knowledge of interdependencies. Some local government agencies had led the development of regional preparedness plans to address infrastructure interdependencies. In most GECCo events, state agencies and the National Guard were directly involved and supportive of regional collaboration.

Most GECCo exercises revealed that much work remains to be done in coordinating local and state government disaster preparedness plans and contingency plans of private sector organizations for a major disaster. Noted during numerous GECCo events was the significant challenge and effort it would take to consolidate and organize the disaster plans across all sectors.

Private sector and other nongovernment organizations emphasized the need for their inclusion in regional preparedness planning, not just with the state, but with local government. It was often noted that public and private sector representatives must be willing to meet and participate in regular emergency management exercises and planning events.

Nearly every GECCo identified the need for more mutual assistance agreements and memorandums of agreement among states, tribes, counties, cities, and with and among private sector organizations. It was also noted that these agreements need to be developed outside the urban areas to include participants in areas outside the potential region impacted by a disaster. In a large-scale event, local mutual aid agreements with neighboring agencies and organizations would not be sufficient.

Consideration for regional and national defense assets in preparedness planning received limited focus other than recognition of the need to incorporate the military into regional preparedness planning. Regional emergency planners, the public and private sectors, the media, and the general public need to understand clearly the limits, delays, and constraints that may affect the immediate receipt of assistance.

4.4.3.3.2 GECCo Recommendations

1. Where possible, response and contingency plans should be shared, coordinated, upgraded, and tested with regional exercises.

2. Undertake efforts by federal, tribal, state, and local governments to improve cooperation and coordination from the bottom up to the national level, including integrating EOCs to facilitate public–private coordination vertically and cross-sector regionally.

3. Maintain an up-to-date list of key stakeholders for points of contact (POCs) responsible for disaster preparedness and management at state, tribal, and local EOCs and be made accessible to all key stakeholders. This should be updated quarterly based on the turnover of staff in many organizations.

4. Incorporate key stakeholder POCs responsible for disaster pre-paredness and management into contact lists with the numbers and emails of their counterparts.

4.4.3.4 Data Sharing

4.4.3.4.1 GECCo Findings

Information sharing was a major focus during each of the GECCo exercises. Most local government, utility, and telecommunications representatives reflected sentiments that private companies are reluctant to share informa-tion directly with government organizations. Through participation in exer-cises like GECCo, however, they can better determine what information is needed and when it needs to be shared. Many utilities noted that a trust relationship is paramount in creating an environment where it is felt that information can be shared safely, and in confidence.

Each GECCo event identified that all phases of emergency management depended on data from a variety of sources. The appropriate data had to be gathered, organized, and displayed logically to determine the size and scope of disaster. During an actual event, it was critical to have the right data, at the right time, displayed logically, to respond and take appropriate action.

Participants often need detailed information concerning communication facilities, buildings, electrical distribution and water systems, and so forth. By using GIS, participants were able to share information more effectively through databases on computer-generated maps. Without this capability, participants had to determine how to gain access to a number of external organizations, and their unique maps and data, and that exercise limited the ability to gather these resources. This resulted in participants having to guess, estimate, or make decisions without adequate information.

Cross-sector information sharing was very challenging, but was acknowl-edged as vital to disaster preparedness and management and response. Electric participants often cited the need to know what the critical loads are for the other sectors, and indicated that without this knowledge, it would be difficult to establish restoration priorities.

4.4.3.4.2 GECCo Recommendations

1. Create an information-sharing working group among key stake-holders to work on approaches and mechanisms to improve infor-mation sharing.

2. Investigate creating a nonprofit organization to serve as the secre-tariat to enable the secure sharing of information and to keep it from public disclosure under state and local "sunshine laws."

3. Leverage GIS to fulfill data requirement needs for planning and emergency operations and become the backbone of emergency management.

4.4.3.5 Risk Assessment and Mitigation

4.4.3.5.1 GECCo Findings

Each GECCo included a tabletop exercise that focused on aspects of risk assessment and mitigation. A majority of participants had indicated their respective organizations had conducted physical and cyber vulnerability assessments and risk assessment approaches to determine where to focus resources. However, most had not undertaken mitigation measures to deal with large-scale disasters. For example, a number of fire departments had not considered an alternative water supply, including temporary above-ground water mains or the ability to pump from freshwater lakes and reservoirs. Their mitigation measures were largely focused on adjusting water use depending on priority water needs.

In looking at interdependencies, many GECCo participants identified transportation as the most critical infrastructure, and not electric power, which was the other candidate for most essential infrastructure. This finding was subjective and derived from the scenario, and not based on a systematic study of the relative importance of particular dependencies. It ultimately varied with what system was affected during the GECCo event, the type of impact, and the conditions under which the disruption occurred.

Similarly, the relative importance of regional critical infrastructures and key resources was specific to the organization, based on participants' knowledge of their infrastructure or organization and, at most, an understanding of only high-level interdependencies. Participants realized the need to look at risk in the context of larger regional disasters, but many participants noted it would require the widening of parameters for disaster response, mitigation, and planning, because such disasters are high-impact, but low-probability events.

4.4.3.5.2 GECCo Recommendations

Develop requirements for and implement a risk-assessment methodology focused on interdependencies and associated physical and cyber vulnerabilities and all hazard threats. The model for this methodology could be developed by DHS in concert with regional stakeholders, and could focus initially on developing criteria to identify and rank critical infrastructure assets and key resources. There was general consensus that developing a standard would be achieved if sectors used a recognized and acceptable methodology.

4.4.3.6 Response

4.4.3.6.1 GECCo Findings

It was largely recognized during the GECCo exercise that the general population in the affected areas would be on their own for days at a minimum, given the level of regional disruptions and outages and the fact that there would be competing needs for federal resources throughout the impacted region.

There is a need for procedures on how and when a backup EOC would be established that would link the state, county, major municipalities, and command centers and EOCs of other key public and private organizations.

Challenges associated with evacuation during the exercises were rarely addressed, and there was low confidence that regional evacuation could move large numbers of individuals from homes and businesses in a chaotic situation of transportation gridlock, no power, and limited communications.

Sheltering large numbers of individuals was acknowledged to be a major problem. Schools would have only a limited amount of food, and many potential shelters could lack heat and potable water, or would soon exhaust available resources. Without electricity, water and sanitary services would likely be a problem.

Certain GECCo events dealt with a large number of casualties that exceeded the surge capacity of hospitals. Utilities and other essential service providers (for example, banks, financial institutions, and hospitals) would be greatly hampered in resuming or maintaining operations because of the inability to bring staff in or to keep personnel from leaving to be with their families. In other instances, organizations would need to shelter individuals who could not return to their homes.

Many participants emphasized the need for a certification process to enable emergency medical, utility maintenance, and key stakeholder essential personnel to have access to buildings and get past roadblocks. It was noted that it could take 2 to 3 days for the National Guard to fully mobilize for the disaster, considering that mobilization would be delayed because of the regional paralysis. Impacts of a major event could be so widespread that the National Guard could be spread thin and sent to high-priority areas.

4.4.3.6.2 GECCo Recommendations

1. Develop a simple credentialing process in concert with DHS and with input from local government officials and private sector and other key stakeholder organizations. This process must also be coordinated with neighboring states to allow critical resources (people and materials) to access restricted areas.

2. Determine what resources are available to sustain first responders (water, food, bathroom facilities, equipment such as blankets, tools, and flashlights.)

3. Identify staging areas and transportation routes to get to the disaster area and assess for potential interdependency-related vulnerabilities.

4. Include community emergency response teams (CERTs) in local emergency planning so they can provide needed depth to first responder activities.

5. Develop contingency plans between local law enforcement, the Federal Bureau of Investigation, and the National Guard to deal with civil unrest.

6. Include local media in exercises and work with them to define their role and how to use their resources for disaster response.

4.4.3.7 Recovery

4.4.3.7.1 GECCo Findings

Most participants did not recognize the extent of recovery and restoration challenges, or how long it would take to remove debris and restore and rebuild structures and critical assets such as electric power transmission and distribution systems. Most organizations appeared prepared for low-level emergencies but not for large-scale disasters.

While mutual assistance agreements were in place (for example, among utilities, local governments, and states), with several GECCos there was no guarantee that they would be honored given a widespread disaster. Most participants agreed their organization would need to be as self-reliant as possible and arrange for mutual aid agreements with organizations outside a particular area that would not be affected by a disaster in the region.

Participants indicated that restoring electric power and water services result-ing from a prolonged outage required cooperation, contingency planning, and exercise and training among regional power companies and government util-ity organizations. The availability of transportation infrastructure would be necessary for restoration of critical infrastructure operations and other essential services. Impediments to travel could be compensated by use of marine transportation, or medium and heavy lift helicopters, if such assets are available.

The security of infrastructures during the restoration process was also identified as a concern. There would be a need to protect critical assets and resources such as fuel, power generators, and other equipment.

Each GECCo event included discussion on priorities regarding service restoration in an environment when there would be great demand and competition. Most participants pointed out that states, localities, and utili-ties had already established priority lists, and these should be followed. All participants noted that priority restoration should be flexible, depending on need. At the same time, most participants appeared to understand that in a major disaster, priority lists would likely change, and infrastructure interdependencies should play a role in which services were restored and in what sequence.

The way to manage the influx of volunteer aid, including people, food, clothing, materials, and equipment, from outside the region was not appar-ent or well addressed by GECCo participants. Also unclear was what organi-zation would be in charge of managing such donations or how organizations or jurisdictions that needed these resources would be identified and prioritized according to criticality of need, or how the donated service of materials would be dispatched to where it was most needed.

4.4.3.7.2 GECCo Recommendations

1. Develop a clearinghouse for resources management to enable providers and requestors to register their respective supplies, products, services, and needs.

2. Implement the U.S. National Grid (USNG) (discussed in detail in Section 8.3) to support recovery activities. This would greatly enhance coordination among search and recovery teams.

3. Develop a cooperative long-term regional recovery restoration strategy that takes into account all key stakeholder interests and which recognizes that the postdisaster status of the impacted communities will be different than that preevent.

4. Establish criteria and a plan for conducting system and structural certification inspections as part of disaster preparedness. This includes the development of a debris management plan.

5. Determine the need for out-of-region workers and develop a plan for accessing, certifying, and bringing in personnel resources from outside the area if required.

6. Create procedures to enable businesses to contribute resources without fear of liability. This includes the development of *Good Samaritan* laws to facilitate volunteer assistance.

7. Hold community workshops that focus on what both civilian and defense federal authorities can contribute in terms of services and resources for recovery and restoration, including the examination of issues associated with access to these services and resources and their effectiveness, including impediments, and recommend ways for improvement.

8. Develop state and local government and regional military facility guidelines to use vessels to transport basic necessities and essential components and equipment to areas that may be impassable to land transportation.

4.4.3.8 Emergency Management Responsibilities

4.4.3.8.1 GECCo Findings

Among all the issues explored during the GECCo exercises, none was more challenging than addressing the question of who was in charge, and the related problems associated with sorting out organizational roles and responsibilities in a major disaster when these roles are changing through the life cycle of a disaster.

Many participants questioned that ability of government to develop a clear line of authority quickly among agencies at different levels and across jurisdictions. A common theme was that leadership is vital during a disaster, and that the quality of those with senior responsibilities and their ability to share in decision making on priorities will determine how well disaster response is executed.

It was not uncommon that participants either were not familiar with the National Response Plan or did not believe that it would function or be executed as written. Both public and private organizations alike observed that local and state plans are not written to complement the NRP, and vice versa.

Representatives from civilian entities, state and local governments, and federal government agencies cited the NRP and the National Incident Management System (NIMS) as the solution to the roles and missions issue. They saw it as only a matter of training state, local, and private sector stakeholders in NRP and NIMS procedures.

Most GECCo participants were in agreement on the need for more and ongoing NRP and NIMS training. It was often noted that local jurisdictions, utilities, businesses, and other organizations have their own disaster response or contingency plans and responsibilities to employees, customers, and in the case of corporations, their shareholders. Participants often raised the importance of ensuring that plans are flexible guidelines and do not impede response and recovery with bureaucratic or legal obstacles.

Government participants often pointed to the fact that the state was in charge and would call in the federal government when state resources and the National Guard were overwhelmed. Other participants observed that the catastrophic nature of an event such as an earthquake or hurricane would spur the federal government to take action nearly immediately without waiting for the formal process to take place for the president to declare a national disaster.

It was not uncommon to find that the military and military assets would be incorporated into the response and recovery. How military assets would be deployed, used, and participate in response and recovery activities was never clear, though, which raised the question and importance of the military's involvement in major disasters.

4.4.3.8.2 GECCo Recommendations

1. Hold regional interactive workshops focused on determining who would be in charge to understand the specific roles and responsibilities, and mission responsibilities and associated challenges.

2. Create a working group to delineate roles and missions, thereby leveraging existing federal, tribal, state, and local response plans and knowledge of response, recovery, and restoration needs from lessons learned.

3. Conduct incident management procedures training for key stake-holders and hold regional and targeted exercises to work through chain of command issues.

4.4.3.9 Business Continuity and Logistics

4.4.3.9.1 GECCo Findings

It was noted by many government participants that businesses such as retail, manufacturing, distribution, and service organizations are rarely involved directly in local or regional preparedness planning. Most participating businesses and organizations, with the exception of larger companies, had neither the time nor the personnel to focus on disaster response planning. They are inward focused and generally do not interface with government or other organizations on preparedness issues. Long-term restoration would be dependent on residents remaining in, and or returning, to the region, and government assistance would be needed for financial assistance and other incentives, particularly for small businesses in the disaster area.

Organizations need to recognize that an existing contract for critical services, supplies, and equipment may not be valid if another organization holds a similar contract for the same type of assistance during emergencies. In a prolonged infrastructure disruption, maintaining integrity of the food supply, which is highly dependent on power, clean water, waste treatment, refrigeration, and transportation, is essential.

Participants indicated that their organizations do not pay enough attention to "people issues" in their contingency planning and need to find ways to ensure that essential personnel are provided incentives, including assurances that their families will remain safe.

4.4.3.9.2 GECCo Recommendations

1. Encourage all organizations to examine and reassess their contingency plans based on the findings and recommendations in their respective GECCo report and lessons learned.

2. Create an internal incident management structure and guidelines for organization staff to follow in a major disaster. Organizations should put in place procedures to ensure that they have identified all essential personnel that would be required to support the business or government agency in a major disaster.

3. Government agencies and utilities should investigate digitizing and backing up important system information outside the geographic area to a site or sites that would not be impacted by earthquake or other disasters striking their facilities.

4. Organizations should investigate designating a single location (alternate site) with sufficient resilience; they should locate an area

or facility outside the region from which to conduct business in a major disaster.

5. Develop and share cooperative arrangements for use with key suppliers and customers that enable assessment of cost-effective security and resiliency needs for supply chains.

6. Organizations should identify critical suppliers, products, and material and work with their suppliers to identify and assess supply chain vulnerabilities/interdependencies and disruption impacts.

7. Work with key suppliers on interdependencies and conduct on-site assessments that focus on critical services (for example, energy and water systems) and establish high-order priorities for risk reduction.

4.4.3.10 International GECCo

The GECCo program has become known across the international GIS and emergency management communities and resulted in discussions with the United Nations (UN). As a result, the UN Office of Outer Space Affairs, in collaboration with GITA, conducted a GECCo event for the country of Vietnam in 2014. This was part of a larger program to support ongoing UN efforts to support disaster management in this country.

The emphasis of the Vietnam GECCo was threefold. The first objective was to demonstrate and provide the national government with a framework for enabling spatial collaboration at the local and regional levels across the normally disconnected elements that must work together during a disaster. The methods and means by which the national government could then replicate such workshops nationwide, reasoning that the types of organizations and the threats faced vary geographically across Vietnam, could be provided to key workshop participants. In this way, the government is empowered with tools to build capacity at the local and regional levels before, during, and after a disaster.

The second objective of the workshop was to foster the personal relationships and knowledge required at the local and regional levels for successful collaboration during a disaster. Prior experience indicated that a lack of technical ability or spatial data is seldom a cause for failing to use GIS-based technologies effectively during a crisis. Rather, the root causes of such a failure more often are (1) a lack of awareness about who to contact to obtain required data, (2) a poor understanding of what spatial products and services are needed to support an event, and (3) an absence of mechanisms by which data may be shared quickly and effectively.

The outcomes for this objective included the establishment of professional relationships whereby such issues can be discussed on an ongoing basis and initial areas of successful models of collaboration across organizations are identified and championed. They also included the creation of an increased cadre of individuals trained in the GECCo method to help propagate the program of work nationally.

The final and overarching objective of this workshop was to help facilitate the use of space-based imagery available through means such as the international charter and Sentinel Asia at the local and regional level during a crisis. Previous UN studies clearly documented that Vietnam understands the importance and usefulness of such GIS-based technologies during a disaster.

However, those studies also identified areas of potential improvement with respect to turning the spatial data into actionable intelligence products useful to responders and decision makers working at the actual disaster site. The 2014 GECCo in Vietnam resulted in the identification of a series of recommendations to support a framework for enabling spatial collaboration at the local and regional levels across the normally disconnected elements, which must work together during a disaster.

4.5 Howard Street Tunnel Disaster

The issues of data sharing and infrastructure interdependence that the GECCo approach was designed to highlight were demonstrated and confirmed in one widely publicized disaster in 2001. At approximately 1500 on the afternoon of July 18, a 60-car freight train operated by CSX entered the Howard Street Tunnel near downtown Baltimore, Maryland. It derailed and several cars containing hazardous materials caught fire.

In the official National Transportation Safety Board (NTSB) report (USDOT, 2002), CSX stated that Baltimore City Fire Department was notified of the incident at 1535, whereas official fire department records indicate that the alarm was given at 1615. Anecdotal reports suggest that the fire department may have been called earlier for reports of smoke coming from manholes, but that the source of the fire could not be located.

Regardless of the alarm time, the fire soon developed into a massive blaze that issued forth tremendous volumes of dense, black smoke and choked the tunnel with heat in excess of what could be safely navigated by fire crews. These conditions prevented firefighters from being able to set up remotely operated hose lines to combat the fire because its location could not be readily determined in the 2,800-m long tunnel. By 1713, Baltimore City Fire had struck 5 alarms, bringing a total of 17 engines, 7 ladders, a rescue company, and a host of chiefs, medics, and support personnel to the scene (USFA, 2001).

Among the many concerns of the fire department, given the train's bill of lading, which indicated numerous tank cars filled with flammable liquids, including tripropylene, was the potential for a boiling liquid-expanding vapor explosion (BLEVE). A BLEVE occurs when a container, in this case, a railroad tank car, fails and explodes with tremendous force due to increasing vapor pressure resulting from the liquid inside boiling. The potential for a BLEVE is most severe when the containment vessel is only partially filled because the potential for large volumes of flammable gases under extreme pressures is substantially greater than that found in vessels that are completely full.

At approximately 1815, a 1-m diameter water main ruptured at the intersection of Howard and Lombard Streets in the immediate vicinity of the fire. In concert with the Baltimore Public Works Department, it was decided to let the main flow freely for a period of at least 2 h. As it did so, the color of the smoke changed, indicating the production of steam and the likelihood that the leaking water was reaching the fire.

The water also damaged underground power cables, causing approximately 1,200 buildings to go without power for days (FRA, 2005). The Maryland Department of Environment was summoned to the scene and tested the smoke, confirming the presence of large amounts of water vapor. In all, a total of approximately 2.26 billion L of water escaped from the broken main before this leak was shut off at 2300.

While this averted a BLEVE, the fire continued to burn uncontrolled for days, despite the lowering of a large-diameter hose line into the tunnel through a manhole above the fire. By 0900 Thursday, nearly 3 days after the incident began, firefighters were able to lower the temperature in the tunnel from more than 800°C to one low enough to allow the remaining fire to be brought under enough control for the smoldering cars to be dragged from the tunnel. The fire was not completely extinguished until late in the day on July 22, and the tunnel was reopened on July 24, 2001.

The following is a partial list of additional damages and disruptions caused by the derailment and fire:

- Train service to the Port of Baltimore was severely limited. This resulted in a near stoppage in the flow of containers and motor vehicles to and from the port, the second busiest in these categories on the East Coast.
- The Inner Harbor for the Port of Baltimore was closed for more than 24 h.
- Mail and international shipping interests were disrupted for as long as 3 weeks.
- The water to the Shock Trauma Center and several other critical businesses in the area had to be supplied from alternative sources or trucked in. It took 12 days to repair the broken main.
- The storm drainage system in the area collapsed and failed, resulting in water damage to many area businesses and homes.
- Traffic was severely disrupted along numerous main traffic arteries for weeks. All major arteries entering the city of Baltimore were closed at the onset of the fire.

- Major League Baseball canceled a series of games to be played by the Baltimore Orioles. This alone resulted in an economic loss of nearly $4.5 million.
- The fiber-optic backbones of seven of the largest U.S. Internet service providers (ISPs) were damaged and communications with critical interests abroad were severed or significantly hampered, including the U.S. embassy in Zambia. More than 9.1 km of fiber-optic cable had to be replaced.
- Two tunnel workers from CSX were injured and hospitalized.

Despite the thousands of pages written in both official and news reports, none mention the potential for GIS to influence any aspect of the emergency management life cycle had it been applied to this event. GIS would have proved useful in understanding the potential for cascading infrastructure failures. While speculative, it is reasonable to assume that had the realization about the potential for a disaster existed, the presence of coincident geometry associated with so many infrastructure elements would have resulted in mitigating actions.

Similarly, the integrated use of GIS might have proven especially helpful in predicting failures when the event spiraled out of control, provided better information about access and egress to responders earlier in the event, and aided with the recovery process. Then again, perhaps not, as the *Baltimore Sun* reports that little has changed since the fire (Dresser, 2011). In fact, another massive derailment occurred on August 6, 2010, when 13 cars in a 79-car train left the tracks. This incident underscores the importance of data sharing among infrastructure stakeholders and the importance of using geospatial technologies as a tool for critical infrastructure management.

References

Austin, R.F. (2005). *A Community Framework for Critical Infrastructure Protection: The GECCo Initiative*. Homeland Infrastructure Foundation-Level Database (HIFLD) Working Group, Washington, DC.

Austin, R.F. (2010). *The Geospatial Information and Technology Association's GECCo Initiative—Status and Update*. Homeland Infrastructure Foundation-Level Database (HIFLD) Working Group, Washington, DC.

Austin, R.F. (2012). Guidelines for geospatial data sharing. Presented at Geospatial World Forum Proceedings, Amsterdam.

Austin, R.F., Sherin, J., Driggers, R., and Watson, W. (2010). Panel: Bird's eye view: The critical role of geospatial information to protect our nation's infrastructure. Presented at 2010 CIP Conference: Manage Risk with Resilience, Washington, DC.

Bush, G.W. (2003). Critical infrastructure identification, prioritization, and protection. Homeland Security Presidential Directive/HSPD 7. December 17. http://georgewbush-whitehouse.archives.gov/news/releases/2003/12/20031217-5.html.

Department of Homeland Security. (2014). National Infrastructure Protection Plan. http://www.dhs.gov/national-infrastructure-protection-plan.

Dresser, M. (2011). 10 years after Baltimore tunnel fire, much is unchanged. *Baltimore Sun*, July 16.

Federal Emergency Management Agency. (2010). Strategic Foresight Initiative: Critical infrastructure long-term trends and drivers and their implications for emergency management. http://www.fema.gov/pdf/about/programs/oppa/critical_infrastructure_paper.pdf.

Federal Railroad Administration (FRA). (2005). Report to Congress: Baltimore's railroad network: Challenges and alternatives. Washington, DC.

Gomez, J.P. (2008). Facilitating a dialog between the NSDI and utility companies. Presented at the Geospatial Information and Technology Association Emergency Response Symposium. Seattle, WA.

Government Accountability Office. (2007). *Critical Infrastructure: Sector Plans Complete and Sector Councils Evolving*. GAO-07-1075T. July 12. www.gao.gov/assets/120/117283.pdf.

Government Accountability Office. (2009). *The Department of Homeland Security's (DHS) Critical Infrastructure Protection Cost-Benefit Report*. GAO-09-654R. June 26. www.gao.gov/new.items/d09654r.pdf.

Government Accountability Office. (2013). *DHS Needs to Improve Its Risk Assessments and Outreach for Chemical Facilities*. GAO-13-801T. August 1. www.gao.gov/assets/660/656482.pdf.

Obama, B. (2013). Critical infrastructure security and resilience. Presidential Policy Directive/PPD-21. February 12. http://fas.org/irp/offdocs/ppd/ppd-21.pdf.

U.S. Department of Transportation (USDOT). (2002). Effects of catastrophic event on transportation system management and operations: Baltimore, MD—Howard Street Tunnel fire—July 18, 2001. Washington, DC.

U.S. Fire Administration (USFA). (2001). "CSX Tunnel Fire, Baltimore, Maryland." Technical Report USFA-TR-140. Department of Homeland Security, U.S. Fire Administration, National Fire Data Center, Emmitsburg, MD.

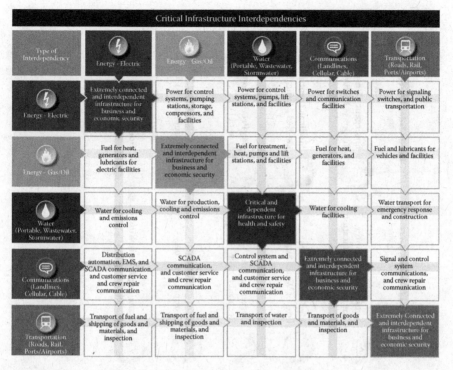

Type of Interdependency	Energy - Electric	Energy - Gas/Oil	Water (Portable, Wastewater, Stormwater)	Communications (Landlines, Cellular, Cable)	Transportation (Roads, Rail, Ports/Airports)
Energy - Electric	Extremely connected and interdependent infrastructure for business and economic security	Power for control systems, pumping stations, storage, compressors, and facilities	Power for control systems, pumps, lift stations, and facilities	Power for switches and communication facilities	Power for signaling switches, and public transportation
Energy - Gas/Oil	Fuel for heat, generators and lubricants for electric facilities	Extremely connected and interdependent infrastructure for business and economic security	Fuel for treatment, heat, pumps and lift stations, and facilities	Fuel for heat, generators, and facilities	Fuel and lubricants for vehicles and facilities
Water (Portable, Wastewater, Stormwater)	Water for cooling and emissions control	Water for production, cooling and emissions control	Critical and dependent infrastructure for health and safety	Water for cooling facilities	Water transport for emergency response and construction
Communications (Landlines, Cellular, Cable)	Distribution automation, EMS, and SCADA communication, and customer service and crew repair communication	SCADA communication, and customer service and crew repair communication	Control system and SCADA communication, and customer service and crew repair communication	Extremely connected and interdependent infrastructure for business and economic security	Signal and control system communications, and crew repair communication
Transportation (Roads, Rail, Ports/Airports)	Transport of fuel and shipping of goods and materials, and inspection	Transport of fuel and shipping of goods and materials, and inspection	Transport of water and inspection	Transport of goods and materials, and inspection	Extremely Connected and interdependent infrastructure for business and economic security

FIGURE 1.2

Interdependency relationships. (*Source*: Geospatial Information and Technology Association, Birds of a Feather Committee's Report on Critical Infrastructure Interdependencies, 2008.)

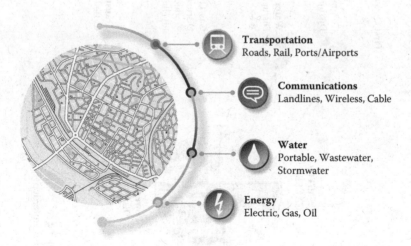

Transportation
Roads, Rail, Ports/Airports

Communications
Landlines, Wireless, Cable

Water
Portable, Wastewater, Stormwater

Energy
Electric, Gas, Oil

FIGURE 1.3

Interdependency relationships.

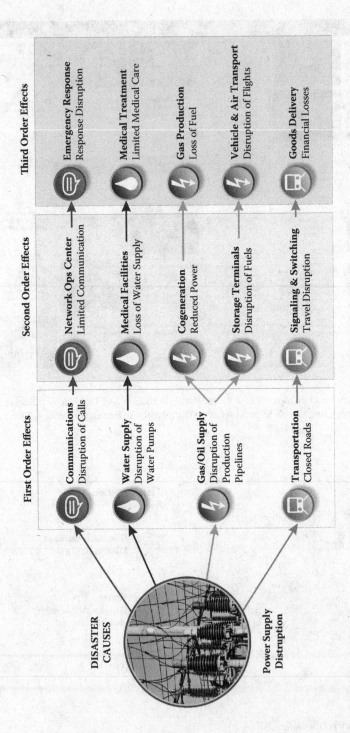

FIGURE 1.4
Cascading interdependencies.

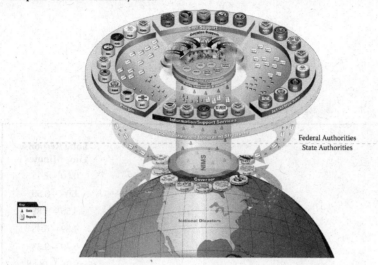

FIGURE 3.2
Homeland Security (HLS) Geospatial Concept of Operations (GeoCONOPS). (From Homeland
Security Geospatial Concept of Operations [GeoCONOPS]. With permission.)

FIGURE 5.3
Wilson hydrants.

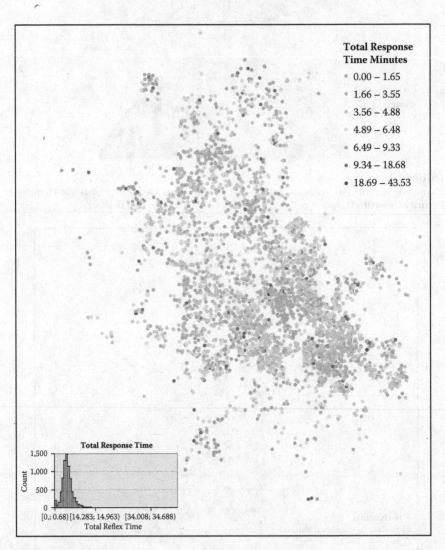

FIGURE 5.4
Total response time.

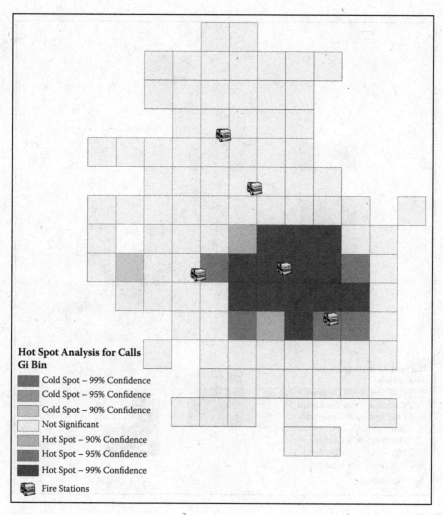

FIGURE 5.5
Hotspot analysis for calls.

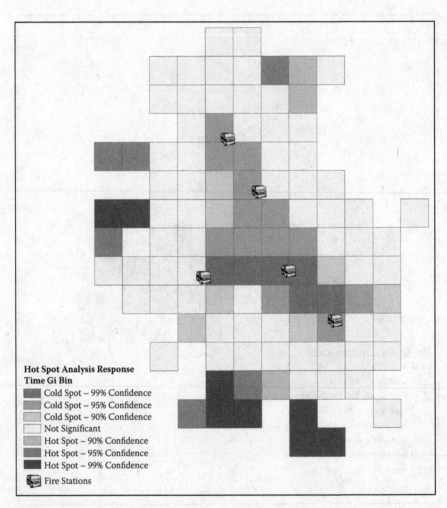

Hot Spot Analysis Response Time Gi Bin
- Cold Spot – 99% Confidence
- Cold Spot – 95% Confidence
- Cold Spot – 90% Confidence
- Not Significant
- Hot Spot – 90% Confidence
- Hot Spot – 95% Confidence
- Hot Spot – 99% Confidence
- Fire Stations

FIGURE 5.6
Response time hotspots.

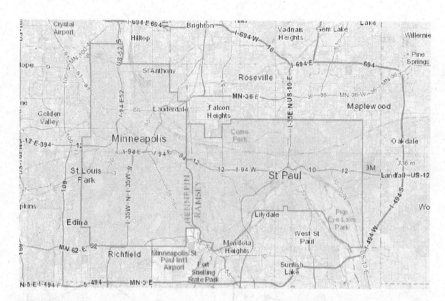

FIGURE 7.1
Twin Cities metropolitan area.

FIGURE 8.1
GIS group at MEMA during Katrina.

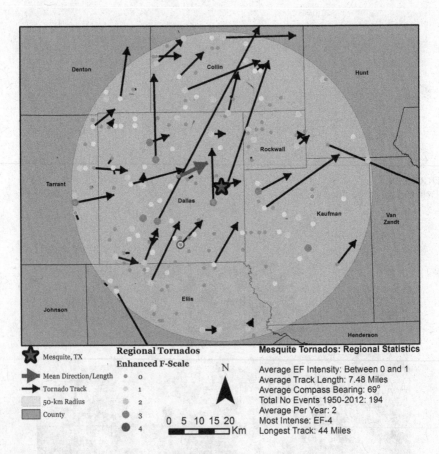

Regional Tornados
Enhanced F-Scale

- Mesquite, TX
- Mean Direction/Length
- Tornado Track
- 50-km Radius
- County

- 0
- 1
- 2
- 3
- 4

N

0 5 10 15 20
Km

Mesquite Tornados: Regional Statistics

Average EF Intensity: Between 0 and 1
Average Track Length: 7.48 Miles
Average Compass Bearing: 69°
Total No Events 1950-2012: 194
Average Per Year: 2
Most Intense: EF-4
Longest Track: 44 Miles

Denton, Collin, Hunt, Rockwall, Tarrant, Dallas, Kaufman, Van Zandt, Johnson, Ellis, Henderson

FIGURE 9.3
Mesquite, Texas, tornadoes (1950–2012).

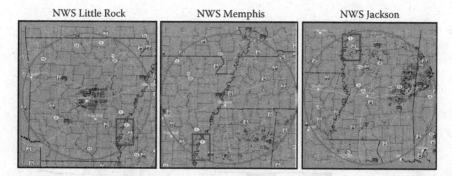

FIGURE 9.7
Effective weather radar coverage for Bolivar County, Mississippi.

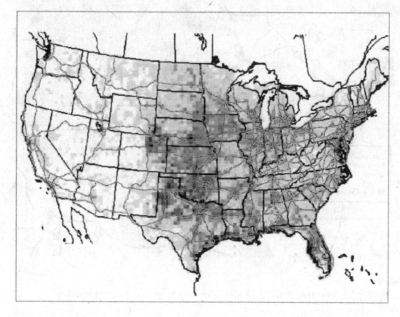

FIGURE 9.9
Thirty-kilometer aggregate tornado data for the United States (1950–2002).

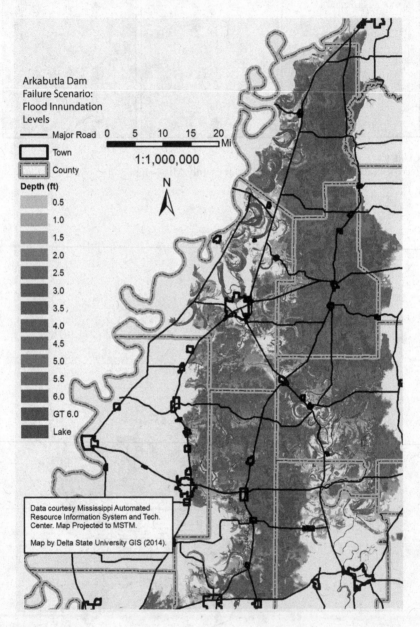

Arkabutla Dam Failure Scenario: Flood Innundation Levels

— Major Road
☐ Town
☐ County

Depth (ft)
0.5
1.0
1.5
2.0
2.5
3.0
3.5
4.0
4.5
5.0
5.5
6.0
GT 6.0
Lake

0 5 10 15 20
Mi
1:1,000,000

N

Data courtesy Mississippi Automated Resource Information System and Tech. Center. Map Projected to MSTM.

Map by Delta State University GIS (2014).

FIGURE 9.12
Flood depth grids reflective of an Aklabutla Lake dam breech due to earthquake.

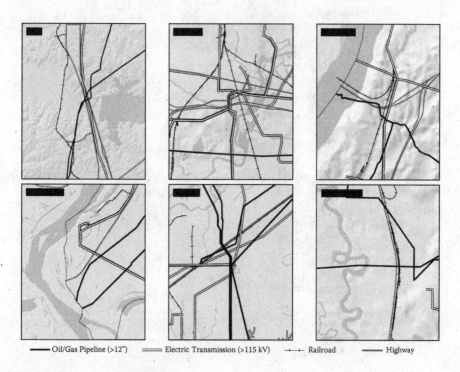

——— Oil/Gas Pipeline (>12")   ═══ Electric Transmission (>115 kV)   ⊢⊣ Railroad   ——— Highway

FIGURE 9.13
Geographic convergence of significant critical infrastructure elements.

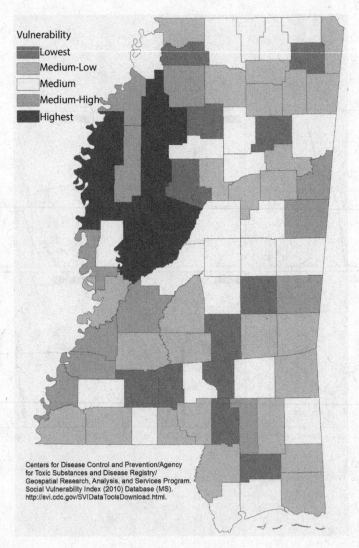

Vulnerability

- Lowest
- Medium-Low
- Medium
- Medium-High
- Highest

FIGURE 9.15
Social Vulnerability Index for Mississippi counties (2006).

5

Local Government Application of GIS to CIP

5.1 Introduction

Emergency response systems are themselves critical infrastructure, but they are also the primary means by which other critical infrastructure are protected. Emergency response assets, when viewed as critical infrastructure, are interdependent with other aspects, such as water and communications systems. Further, because all emergencies or disasters are local in origin, nearly all police, fire, and EMS agencies are managed and operated by local government.

5.2 First Responder Computer-Assisted Dispatch Systems

Emergency telephone number systems are in wide use globally, and their objective is to provide a standardized telephone number that may be dialed in the event of an emergency. Within the United States and Canada, this number is 911. A basic 911 system is designed such that a set of telephone number prefixes are automatically routed from their respective switching stations to a public service access point (PSAP) when 911 is dialed.

The degree of complexity and integration with other technologies, such as mapping and radio systems, varies greatly from the most basic system, which simply routes calls to the PSAP, to those that integrate global positioning system (GPS), radio, and mapping systems into a common system using computer-aided dispatch (CAD). The majority of systems in place within the United States and Canada use Enhanced-911, often referred to as E-911. In the United States, E-911 systems are governed primarily by regulations and guidance managed by the Department of Transportation (DOT), the Federal Communications Commission (FCC), and the National Emergency Number Association (NENA).

E-911 systems are designed to enable dispatchers with the ability to determine the location of a caller—the caller's point of presence (POP)—even if the caller becomes disconnected due to violence or other circumstances resulting in a loss of connection. Under current guidance, E-911 systems must achieve this for both wire-line and cell phone–based calls. In the instance of wire-line calls, telephone numbers are matched to street address and entered into a Master Street Address Guide (MSAG) in database format.

Caller ID technology is present in the PSAP switching system, called the automatic number identification/automatic location identifier (ANI/ALI). Each entry in the MSAG is coded with a third piece of information, often based upon telephone prefix boundaries, called the emergency service number (ESN).

Each ESN is associated, either manually or within CAD/GIS (geographic information systems) software, with an appropriate emergency service provider's jurisdiction or what is sometimes referred to as the emergency service zone (ESZ). Within metropolitan or major suburban areas, the ESN may be broken down to match police beat or fire alarm boxes, which are used to delineate the appropriate running orders, or order of response based upon immediate need or projected incident severity for emergency responders.

For wire-line-based calls, the E-911 process is as follows:

1. The POP call is routed to the ANI/ALI at the appropriate PSAP having jurisdiction.

2. The address is retrieved from the ANI/ALI with ESN and provided to the call taker's console position and its supporting software.

3. The address is passed to a mapping module within the console for geocoding and placement on a map display, while the appropriate response is selected based upon logic rules applied to information entered by the call taker against the ESZ/ESN.

Figure 5.1 shows the Phoenix Fire Department E-911 call center in Phoenix, Arizona. Incident information is routed from a primary call taker who classifies the incident as being a police, fire, or emergency medical event. The call is quickly routed to a call taker (foreground) who is provided with location data from the ANI/ALI and who captures additional information. Once the call taker is confident that the basic quantity and value of data required to dispatch appropriate resources have been captured, the data are queued for dispatch by a dispatcher (background).

The CAD system matches the closest available unit to the call disposition information and assigns appropriate units, which are then alerted by both voice and data streams. The map at the front of the room provides a continuously updated view of vehicle location and status. These data are subsequently viewable on mobile data terminals (MDTs) mounted in vehicles (Figure 5.2).

Underlying the efficacy of this system is the need for high-quality address data. Most E-911 systems will perform an initial geocoding pass and attempt

FIGURE 5.1
Phoenix Fire Department 911 call center.

FIGURE 5.2
MDT in Phoenix Ladder Company 1.

to match the address transferred from the MSAG against a point address layer within the GIS. Lacking the availability of such a layer, as is common in many jurisdictions, the address will be geocoded against the best available addressable street segment data set stored in the E-911 database.

The accuracy of the results available for the latter is highly variable and somewhat unpredictable, as address placement is based upon distance along a roadway segment from an intersection. Therefore, both distance from an intersection and the accuracy at which the address was originally assigned determine the quality of the geocoded location.

There are other problems typically associated with the geocoding process. These include duplicate street names within a jurisdiction, incorporation of directional words (for example, east and southwest) into the street name instead of using prefixes, and mismatches between the MSAG and geodata. Such issues are commonplace and degrade the quality of the result. Such errors may result in tragedy as responders are sent to wrong locations.

Within the context of cell phone–based calls, the cell tower network will capture coordinates from the device being used to make the call, if available, or the tower network will functionally perform a range and bearing calculation. This methodology is termed Non-Call-Path Associated Signaling (NCAS), which essentially treats location as a dynamic value within the ANI/ALI system. It is associated with the mobile positioning center (MPC), a system based on American National Standards Institute (ANSI) standards, which in turn is tied to the position-determining entity (PDE) and uses the E2/V2 interface protocol between the MPC and the ANI/ALI to derive an approximate location.

Thus, in the case of a cell phone call to an E-911 system, a coordinate is used in lieu of an address for the determination of location, and supporting mapping systems are equipped to handle either type of placement of the POP and subsequent selection of the ESZ/ESN. A similar pseudo-ANI/ALI approach is used in the instance of the Voice-over-Internet Protocol (VoIP), but with the introduction of network location affiliated with the POP and, again, the use of the E2/V2 protocol for bridging.

More advanced systems integrate additional components such as MDTs, which allow the transfer of incident data, including location, to a display device mounted in vehicles or carried by the responder. Recreational-grade GPS receivers are often tied to an MDT, thereby allowing the responder to view the location with respect to that of the reported incident.

Jurisdictions, such as those encompassed by the metropolitan Phoenix, Arizona, area, add another layer of sophistication by improving GPS receiver quality with supplementary equipment. These enable the collection of higher-level precision location information, which is shared with the PSAP and other responders.

This not only facilitates the coordination of resources and needed command and control elements associated with larger responses, but also permits the dispatch of the closest available units in place of the standardized running orders set forward by the ESZ/ESN approach. Lastly, *Next Generation 911*, or NextGen 911, promises to break completely free from reliance upon ESZ/ESN by using location-based queries to assemble responses based upon the closest suitable resources within a GIS-based environment.

The lack of integration and common level of implementation among PSAPs presents a significant challenge to the advancement of emergency number systems. Additional concerns stem from the required encoding and securing of data transmitted wirelessly and the amount of bandwidth available to do so. Public safety organizations are discipline specific (for example, police, fire, and emergency medical services). Therefore, requirements and standards related to mobile data transfer are most often developed independently of each other.

While law enforcement has embraced the National Information Exchange Model (NIEM) standards for encoding and transmitting data, the fire service is developing a different standard per the National Fire Protection Association (NFPA). This is to the detriment of advancing emergency number systems, as it creates an additional burden that compounds the lack of consistent levels associated with the technology deployment.

Neighboring PSAPs with differing levels of capabilities often find it difficult to transfer calls from one center to another, a task often required when caller location, as determined by a wireless VoIP network, is uncertain. The call may be routed to the nearest PSAP with respect to cell phone tower location, but that receiving PSAP may not have jurisdiction. This fact is often not uncovered until well into the call taker's interview process and necessitates the transfer of the call to the correct jurisdiction.

In some instances, the transferring agency lacks the ability to transfer location data with the call; in others, the receiving PSAP system may be unable to decode the information transferred to it from a foreign system. This, combined with the increasing costs associated with operating 911 systems, is leading toward a reduction in the overall number of PSAPs while greatly expanding the geographic region to which they provide coverage.

As with geocoding addresses from an MSAG against a wire-line-based call, there are challenges inherent in the use of location from cell and VoIP-based calls to 911. An informal study of all fire calls within the United States indicated that approximately one-third of all requests for help did not occur at a street address. Further, the overwhelming majority of these calls were associated with various elements of critical infrastructure.

The most common types of incidents were associated with automobile accidents along remote stretches of roadways. However, calls involving rail- and waterway-based transport systems, wildland-based events such as fires and missing persons, and electric, gas, and water utilities were certainly not uncommon. The challenge here is that although a 911 call taker may receive multiple calls reporting a gas pipeline failure or other catastrophes and may see the location clearly on his or her map display, the act of communicating that location to responders is problematic because no street address exists for the incident location.

Infrastructure stakeholders, particularly those managing utilities, roadways, waterways, and railroads, typically identify locations using linear referencing systems (for example, mile marker). These locations are particularly difficult to identify and access in the field unless a responder has intimate knowledge of the area in question. As previously identified, during major events, responders often lack that sort of detailed knowledge and are left lost.

5.3 Improving the Standard of Cover for Emergency Services

The accreditation of emergency response organizations is gaining popularity as a means of building public confidence in safety organizations, ensuring continued self-improvement within said agencies, and assisting with national continuity and interoperability within the emergency services sector. As with most accreditation processes, those in emergency services are administered by professional societies or organizations, are granted for set time periods, and are granted largely based upon a combination of metrics that evaluate performance and compliance and the plans to improve them continually. Understanding a jurisdiction's risk and its potential to change temporally is thereby central to improving a public safety or emergency response organization's ability to successfully plan for, mitigate, respond to, and recover from an emergency of any sort.

The results of such introspective looks at risk and a public safety organization's ability to deliver services are known as a standard of cover (SOC) or a deployment analysis. While this section will illustrate the standard of cover process with respect to the fire service, the fundamentals apply to any organization responsible for responding to an emergency. At the outset, a standard of cover seeks to establish the following using a detailed and well-supported data-driven approach based upon the results of a risk analysis, prior performance (critical task analysis), and prior outcomes:

1. The minimum number and types of resources required to respond to common types of emergency events, which ensures adequate safety for both the community served and all responders

2. The basis for a strategic plan that supports continued improvement as measured through both well-defined performance objectives and their underlying meaning with respect to tactics

3. The basis for organizational policies and a defensible position related to adverse events and the criteria used to make decisions (Oregon State Police–Oregon Office of the State Fire Marshall, 2014).

This work, similar in many ways to the hazard mitigation planning process, is best accomplished using a team having significant authority to conduct needed tasks. The importance of a thorough risk assessment, critical task analysis, and review of prior outcomes cannot be underscored enough. A multidisciplinary approach will render the most useful results. The introductory risk assessment and hazard mitigation process outlined in Chapter 1 provided examples associated with an assessment of natural hazards.

Additional components must be considered in the preparation of a risk assessment for use in developing a standard of cover. These include the following:

1. A community risk assessment with respect to the emergency services discipline for which the standard of cover shall be created. For example, a fire department may evaluate each structure or a statistically significant sample thereof to generate an Occupancy Vulnerability Assessment Profile (OVAP) score that corresponds to the relative risk of fire in a particular building. A high-value, high-occupancy structure would score far higher than one that was low value and unoccupied.

2. A technological hazards assessment should be prepared. This should include points of vulnerability with respect to where systems failures, data errors, and cyber attacks would significantly hamper an emergency response or create the need for one in their own right.

3. A human hazards risk assessment should be prepared. This underpins the SOC's understanding of potential harms caused by terrorism, hazardous materials incidents, or biological/disease outbreaks, such as flu pandemic. It also speaks to the potential for everyday human errors and accidents to cascade out of control.

4. A security hazard assessment should be performed to supplement the security risk assessment. Security hazards estimates are concerned primarily with unauthorized persons gaining access to or control of restricted use areas. A security risk assessment must be as concerned with password management policies as it is with the physical security of spaces and places.

The planning team must also identify critical tasks associated with the emergency response process. This requires data about prior performance of such detail and completeness to render useful information. The team may discover that such information is lacking and, consequently, should recommend policies and procedures whereby improved data may be obtained.

For example, NFPA 1710, "Standard for the Organization and Deployment of Fire Suppression Operations, Emergency Medical Operations, and Special Operations to the Public by Career Fire Departments" (NFPA, 2010), contains the metrics applicable to event timekeeping. The standard discerns among alarm answering time, alarm handling time, alarm processing time, travel time, turnout time, and initiating action/intervention time.

Many jurisdictions do not track all of these elements, largely because they lack the personnel or systems required to do so. The SOC team must determine which data elements are most suitable for use in their critical task analysis and determine a means by which both the task and the metric quality may be improved.

Critical task analysis is daunting, but not always quite as challenging as thought at first glance, as GIS may prove of tremendous benefit for analyzing such elements. For example, fire departments are especially concerned with the availability of water for use in fighting a fire. A simple map of hydrant flow rates may provide a reasonable and quick starting place for developing a standard of cover.

Figure 5.3 illustrates the use of GIS to color-code the hydrant flow rates according to NFPA 291 for Wilson, North Carolina. This figure readily shows that the available water flow in the downtown and its immediate residential areas is low. This is likely a consequence of these areas being developed early on in the town's history.

Conversely, those areas developed more recently have greater flow rates and are probably attached to larger and better-designed water system networks. This knowledge should affect both the interpretation of critical task data for the amount of time required to establish a sufficient flow of water to a fire in these areas and the resulting standard of cover, as the deployment of resources should be altered accordingly.

Alterations could include the addition of water tenders (trucks with large water tanks) or an increase in the number of engines used to lay hose and

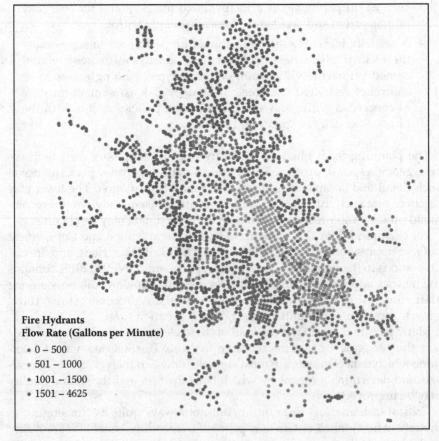

Fire Hydrants
Flow Rate (Gallons per Minute)

- 0 – 500
- 501 – 1000
- 1001 – 1500
- 1501 – 4625

FIGURE 5.3
(See color insert.) Wilson hydrants.

connect to a fire hydrant. Further, the SOC team could perform a cost–benefit analysis and make recommendations pertaining to the upgrading of the municipal water system.

Basic analysis techniques such as this serve as the starting point for more complex analysis. For example, a similar-style map for total response time is difficult to interpret (Figure 5.4). These data are likely partially confounded by outliers that skew the scale, but the sheer density of features makes understanding any underlying trends challenging.

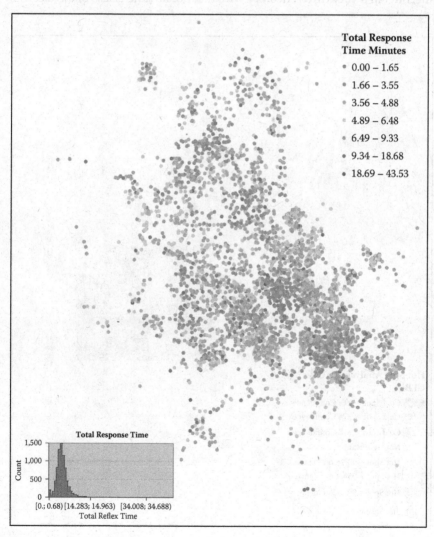

FIGURE 5.4
(See color insert.) Total response time.

These data may be aggregated by total response time, or in this instance, total number of calls for service, into 1-km grid cells that correspond to the U.S. National Grid (USNG) and a hotspot analysis performed. When layered with the location of fire stations, potential problem areas become readily apparent, as it is highly likely that response times in the downtown area suffer as a result of unit availability.

Note that had the aggregation and hotspot analysis been performed using total travel time, the results would appear as shown in Figures 5.5 and 5.6. These results are no less valid than those developed previously for call volume, but tell a very different story—that response time is less of a concern in

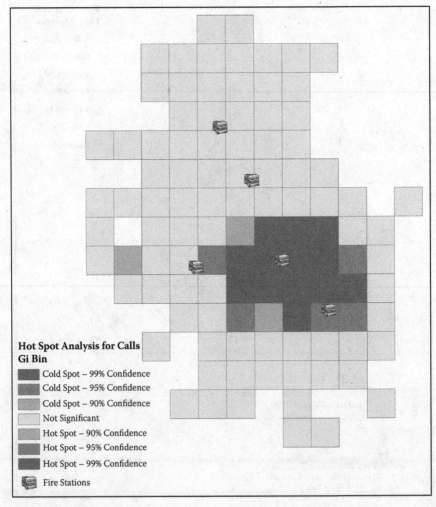

Hot Spot Analysis for Calls Gi Bin

- ■ Cold Spot – 99% Confidence
- ■ Cold Spot – 95% Confidence
- ■ Cold Spot – 90% Confidence
- ☐ Not Significant
- ■ Hot Spot – 90% Confidence
- ■ Hot Spot – 95% Confidence
- ■ Hot Spot – 99% Confidence
- 🚒 Fire Stations

FIGURE 5.5
(See color insert.) Hotspot analysis for calls.

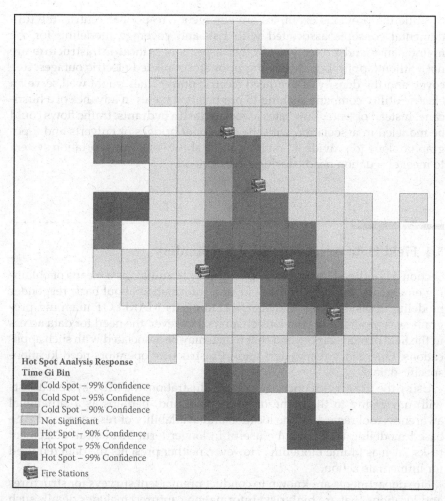

Hot Spot Analysis Response Time Gi Bin
- Cold Spot – 99% Confidence
- Cold Spot – 95% Confidence
- Cold Spot – 90% Confidence
- Not Significant
- Hot Spot – 90% Confidence
- Hot Spot – 95% Confidence
- Hot Spot – 99% Confidence
- Fire Stations

FIGURE 5.6
(See color insert.) Response time hotspots.

high-volume areas than might otherwise have been thought. The objective is to rely upon the expertise of the SOC team to correctly interpret these types of critical task analyses for integration into the SOC document.

From a broader perspective, the SOC team may be able to generate multiple layers of critical task analysis data aggregated using the USNG. Were these data normalized to common data ranges, they might envision a model for predicting the most efficient means of deploying resources as a combination of available water for fighting a fire, the frequency of calls, and total response time. GIS becomes especially valuable as the polygon-based data layers may be converted to raster format and then numerically combined to visualize the interplay among layers to identify where resources may be needed or how they may be more effectively redistributed.

While the previous examples are emergency responder centric, the fundamental concepts associated with risk and coverage modeling for the maintenance and protection of other aspects of critical infrastructure are not without application. Modeling prior storm-related electric outages, tree cover, and the density of overhead electric power lines might well serve an electric utility company seeking to pre-position crews in advance of a hurricane. Instead of water flow rates associated with hydrants, traffic flows could be modeled in association with the density of bridges or culverts and topographic slope to provide a first look at the ability of a transportation system to meet evacuation needs during a rising flood.

5.4 Field Data Access for First Responders

Section 5.1 outlined the use of mobile display terminals as a means of obtaining emergency call data in the field, and information about basic responder modeling tools, such as CAMEO, including its MARPLOT mapping program, was provided in previous chapters. However, the need for data access in the field extends far beyond that which may be associated with such applications. Emergency responders and infrastructure operators need location-specific data.

Using the fire service once again as an illustration, the MDT system assists with navigating to the scene of an incident and, perhaps as a situational awareness tool regarding the location and availability of resources and GIS-based modeling tools, may be useful in longer-term incident management tasks, such as plume modeling. However, neither presents the data required for immediate action.

Fire departments are known to conduct preincident surveys for structures and locations that are high risk. Information captured includes details such as the occupancy load of the building; material stored (such as chemicals); location of exits, standpipes, alarm panels, and Knox-Boxes (where keys are stored); contact information for the building owner and utility service providers; floor plans; and any other relevant information about the facility or structure. These data are often required for rapid, informed decision making, which may make a significant difference in the outcome of an event.

Traditionally, such information is compiled in three-ring binders as paper documents called preplans. However, there is a practical limit to the quantity of information that could be toted around and a reasonable expectation as to whether it could be located by a groggy firefighter bouncing around the cab of a fire truck while en route to a call at 0200.

GIS provides a major improvement to the organization of and access to these types of information. Building floor plans and other location-based data may be displayed as microlayers geo-referenced to a building's footprint, thereby showing the exact location of features of interest. Supporting documentation may be linked to geographic features and retrieved with a touch or a click, providing instant information about not only the presence of things such as hazardous materials, but also response guidelines for dealing with them should they become involved in an incident.

Emerging systems integrate preplan data with systems able to track the location and status of firefighters. Such systems provide a heads-up-style display within the air mask of a firefighter, as well as a live, interactive map for incident commanders. While not yet in wide use, they represent a leap forward in improving the safety of firefighters operating in zero-visibility environments beyond the watchful eye of the incident commander or safety officer.

These systems are often programmed with logic rules that create alerts should a firefighter not move every 60 s or if other firefighter-mounted sensors, such as a thermocouple, report temperatures in excess of a certain maximum threshold. Some systems even facilitate the streaming of thermal imaging video from the firefighter to the incident commander. The latter examples serve to illustrate the merger of remote sensing technologies with GIS in ways not traditionally implemented within the geospatial industry.

With the growth of application areas, thematic elements central to this work once again become limits to the imagination: standards, systems integration, and information management all prove to be the truer boundaries to implementation. Again, the selection of a CAD or MDT system is often based upon that software's ability to meet local requirements and not its ability to integrate with neighboring systems using a standards-based approach.

Systems planners must take note, as this type of lack of interoperability, especially across emergency response disciplines, is often ignored until it appears as a finding in a line-of-duty injury or death report. Such was the case with radio systems interoperability during the catastrophic events of 9/11 (Dwyer et al., 2002), and the problem is likely to continue to occur until a regulatory entity establishes the requirement or funding support becomes tied to standards implementation.

Data management strategies that support critical information access in the field are also especially challenging. Most jurisdictions contain "dead spots" where radio and data traffic are not well received and the use of data streams is impractical.

Conversely, the creation of autonomous systems through the loading of data onto network-independent devices proves difficult from a data maintenance perspective. Data administrators not only must decide when and how to update network-independent devices, but also must ensure that all are updated at the same time to prevent versioning issues, which might endanger property or lives.

As with all technology, carefully laid-out architecture, implementation, and management strategies are keys to success and, in this instance, perhaps survival. While the potential return on investment gained by using geospatial technologies may be quite high with respect to property and lives saved by the responder community, use of these technologies also presents a new and somewhat unknown set of risks to responders as their reliance upon their successful operation continues to rise.

Again, the transportability of mobile concepts is easily applied to other aspects of critical infrastructure management. Many electric utility companies use mobile display terminals to disseminate data and work orders to field crews. Likewise, sharing data via MDT systems to the field crews of two different electric utility companies responding to trouble at a tap is nearly impossible because the requisite standards for doing so are not in place.

5.5 Inventories of Critical Infrastructure

Beyond emergency response and management, the Patriot Act of 2001 defined critical infrastructure as "systems and assets, whether physical or virtual, so vital to the U.S. that the incapacity or destruction of such systems and assets would have a debilitating impact on security, national economic security, national public health or safety, or any combination of those matters." At a federal level, several key categories of critical infrastructure have been identified, as well as the federal agencies responsible for protecting those categories (see Table 1.3).

At the local government level, these broad categories are given specific meaning, and the general principles defined for a nation are brought to life for specific cases. The federal government is responsible for planning coordinated responses to hazards and threats to critical infrastructure. However, state, county, municipal, and tribal governments are responsible for the initial response to such events and work collaboratively with the federal government.

Due to the nature and structure of our government and the need in many instances for security considerations, many activities undertaken by the federal government are defined and shared only within and between government agencies. For example, protective security advisors (PSAs) are assigned by DHS for specific regions throughout the country. These PSAs provide a valuable liaison service between agencies of federal government and local government.

The PSAs also participate in the creation and periodic update of lists of critical infrastructure. Perhaps not too surprisingly, these lists of critical infrastructure are not publicly available. Rather, the lists are used to help frame local response planning, including the construction of GIS support mechanisms.

Most municipal governments maintain detailed information about their infrastructure, critical and noncritical. GIS managers are often called upon to create what is termed a critical asset layer with their GIS database. The

TABLE 5.1

Examples of Critical Infrastructure

• Bridges	• Jails and prisons
• Emergency shelters	• Police stations
• Fire stations	• Storm water facilities
• Fuel sites	• Wastewater facilities
• Hospitals	• Water facilities

critical asset layer is created for the purpose of identifying utility networks, structures, and facilities that would need to be immediately checked after an emergency event. This layer also defines the emergency push routes, which are the roadways, access points, rights-of-way, and trails that must be cleared of debris as quickly as possible to allow first responders to have access to injured people and damaged facilities.

A critical asset layer may be a combination of multiple existing GIS data sets and data maintained in spreadsheets or a work management system by the local government. These source files would typically be reviewed by the specific city utility departments responsible for operations and the criticality of certain assets and facilities confirmed. For example, a wastewater department might operate 143 pumping stations but, based on location and function, only designate 9 of these as critical facilities.

Examples of the types of assets and facilities that might be contained in municipal databases are shown in Table 5.1.

Completing this list of critical infrastructure may also require data sharing with county and state agencies to ensure appropriate collaboration and mutual support. Examples of such data include county roads and state highways, regional water reservoirs, wastewater treatment facilities, public hospitals, and state-managed emergency resources.

These facilities would be geocoded; those located within the municipality or within a reasonable distance from the city limits would be incorporated into the critical asset layer. For obvious reasons, although categories of critical assets can be created for consideration, detailed lists of critical infrastructure are specific to any given city.

There is one glaring omission from this discussion: *privately owned and operated infrastructure*. In Section 4.1, the Strategic Foresight Initiative on the Critical Infrastructure (SFI), among others, noted that "the private sector owns the vast majority of the Nation's critical infrastructure and key resources—roughly 85 percent" (FEMA, 2010, p. 2).

Telecommunications are handled for the most part by publicly held, for-profit corporations. Military bases that operate their own communications centers connect to the adjacent community using connections maintained by those corporations for local communications. City and state governments often lease bandwidth (capacity) from those corporations and are dependent on the availability of that bandwidth.

Petroleum and natural gas pipelines are predominantly owned and operated by publicly held, for-profit corporations. This is also true in the vast majority of cases for electric power, natural gas and propane distribution, fuel oils, and community access television (CATV).

When power is lost, municipal wastewater agencies must hope that their backup generators and fuel supplies will be adequate in the circumstances. An example of this was the adverse effects on the sewage systems of Old San Juan, Puerto Rico, after a loss of power combined with the negative barometric pressure of a passing hurricane.

For-profit corporations have a fiduciary responsibility to their owners and shareholders. Under certain circumstance, data sharing with government agencies could violate that fiduciary responsibility. Consider the case of a telecommunications company that provides information about key underground assets. It is certainly conceivable that a competitor might file a Freedom of Information Act request for that information and also conceivable that the information would be released, to the financial detriment of the facility owner. To facilitate the exchange of such information for critical infrastructure protection and the public good, the government has invoked and elaborated upon the concept of publicly identifiable information (PII).

McCallister et al. (2010), on behalf of the Institute of Standards and Technology, wrote (p. ES1):

> The escalation of security breaches involving personally identifiable information (PII) has contributed to the loss of millions of records over the past few years. Breaches involving PII are hazardous to both individuals and organizations. Individual harms may include identity theft, embarrassment or blackmail. Organizational harms may include a loss of public trust, legal liability or remediation costs. To appropriately protect the confidentiality of PII, organizations should use a risk-based approach; as McGeorge Bundy once stated, "If we guard our toothbrushes and diamonds with equal zeal, we will lose fewer toothbrushes and more diamonds."

On the one hand, for-profit corporations must, due to their fiduciary responsibilities, safeguard PII. On the other hand, those same corporations recognize the need for responsible engagement with the public to ensure the integrity of critical infrastructure, some of which has been built under special legislation designed to permit at least partial monopolies. The conundrum faced by critical infrastructure owners is how to obtain the information they need while protecting the security of that information from public exposure.

The solution has been to create a special method of collecting information about critical infrastructure and key resources (CIKR). Working collaboratively with DHS, for-profit corporations can declare specific information about CIKR to be PII, and thus exempt from public records requests.

Consider the simple example of a buried telephone cable. The public interest may be served by knowing the location of the cable to prevent inadvertent damage to the cable from nearby construction. This is the basis of the "Call before You Dig" services nationwide. However, there is no need for the general public to know the bandwidth (capacity) of the cable, its depreciated capital value, or the composition of the protective sheath surrounding the cable.

Nevertheless, such detailed information could be of use to a city attempting to rebuild and reestablish communications after a disaster. The simple solution is to declare those data PII and exempt from disclosure. The DHS (2014a) report "Critical Infrastructure Sector Partnerships" provides additional information about this concept and its implementation.

In summary, the NIPP, as amended periodically, provides a clear road map for creating an inventory of critical infrastructure for specific government agencies—at all levels of government. To be effective, that inventory should be updated annually to reflect additions, deletions, and changes. In the case of the city of Tampa, staff used the advent of the hurricane season (June 1) as a target date for completing annual updates.

Regular revision is important. Of equal importance is consistency in communications. For example, the DHS risk lexicon was created and is periodically updated to ensure consistency in terminology. The HIFLD Working Group has developed the HSIP data sets, which afford access to consistently defined data to all municipal, county, and state officials charged with critical infrastructure protection.

The FGDC, charged with management of the NSDI, has defined consistent metadata standards to allow systems managed by disparate agencies to be more interoperable. In 2014, the FGDC decided to include the HIFLD Working Group as a subcommittee of the FGDC to ensure broad input and broad access to data.

5.6 Regional Planning and Coordination

Since the terrorist attack on September 11, 2001, and subsequent hurricanes and storms, planning and coordination at the regional level has most notably increased in the Gulf states and northeastern portions of the United States. It has been demonstrated repeatedly that coordination at the local and regional levels is much more effective when dealing with large-scale events.

However, this becomes complicated when federal, state, tribal, regional, and local agencies have different levels of authority when it comes to coordinating emergency management. To make it even more confounding, the official responsibility for emergency planning, response, and recovery is spread among a combination of the different levels of government.

From a federal perspective, regional planning and coordination policies are outlined in the *National Preparedness Guidelines* (NPG) released in 2007. The vision for NPG is, "A nation prepared with coordinated capabilities to prevent, protect against, respond to and recover from all hazards in a way that balances risk with resources and need." (NPG, 2007, p. 1). The NPG was a result of HSPD 8, with state, regional, and local involvement (NPG, 2007, p. iii).

While the DHS identified "expanding regional collaboration as a national priority," it unfortunately has not effectively targeted or supported this priority. As presented in Chapter 4 highlighting the Geospatially Enabling Community Collaboration (GECCo) program, expanding regional collaboration requires the creation and implementation of standard structures, processes, and guidelines across state, regional, and local levels. In the case of the Twin Cities GECCo presented in Chapter 7, this involved the identification of geographic regions that created the capacity to develop and maintain a greater level of coordination among the necessary local, regional, and state agencies.

To achieve meaningful regional collaboration, it is important to keep in mind that all disasters happen locally. Unfortunately, the federal government does not always take this premise into account when it comes to its policies and programs. These policies and programs have conventionally been built from the top down and do not effectively enable a regionalized approach for emergency management and response.

While regional planning and collaboration need to incorporate local jurisdictions and clearly define their roles and processes, local plans among cities and counties must be developed in concert with each other. The end result will be greater regional coordination as witnessed by the findings from the GECCo program.

This was highlighted in a report from the Government Accountability Office (GAO) that reviewed the effectiveness of regions of the United States that are organized locally instead of being imposed by federal or state government, saying they are more likely to have identified a coherent regional area. For example, the federal grant program known as the Urban Areas Security Initiative (UASI) reinforces regional boundaries for enhancing preparedness in high-risk metropolitan areas.

UASI boundaries are determined through a terrorism risk analysis by the most populous metropolitan statistical areas in the United States. This analysis, in turn, was based on the Implementing Recommendations of the 9/11 Commission Act of 2007 (DHS, 2009, p. 25).

The original method for identifying UASI regions was to set a radius around an urban center. While FEMA later changed the method for determining a region, it did not require the addition of any agencies to the governing structure.

FEMA only required the UASI to expand its efforts to include regional partners. The consequence of not including these newly recognized jurisdictions into the region constrains regional collaboration and communication among agencies. One could argue that the grant program whose charge is to improve regional preparedness is actually constraining it.

The Regional Catastrophic Preparedness Grant Program was authorized by Congress in 2007. Its mission is to "support an integrated planning system that provides for regional all-hazards planning for catastrophic events and the development of necessary plans, protocols and procedures to manage a catastrophic event" (DHS, n.d.). This grant focuses on the high-risk UASIs and adjoining regions where its impact will be greatest for increasing regional security and resilience. Unfortunately, this program stands to reinforce the boundaries of the UASI regions, which can widen the gap among those participating agencies in the region and those who are not.

The Homeland Security Grant Program (HSGP) plays a main role in implementing the National Preparedness System by supporting and delivering the core capabilities for achieving the National Preparedness Goal. The National Preparedness Goal objectives indicate that the core capabilities are not exclusive to any single level of government, organization, or community. Rather, they require the combined effort of the whole community.

The HSGP is comprised of three interrelated grant programs:

- State Homeland Security Program (SHSP)
- Urban Areas Security Initiative
- Operation Stonegarden (OPSG)

Together, these grant programs fund a range of preparedness activities, including planning, organization, equipment purchase, training, exercises, and management and administration at the state, regional, tribal, and local levels.

As an example, the HSGP funding priorities for fiscal year 2013 included the following:

- Evolving and enhancing state and major urban area centers
- Implementing a holistic community approach to security and emergency management
- Innovating and sustaining support for the national campaign for preparedness
- Building and sustaining law enforcement terrorism prevention capabilities

The objective of SHSP is to provide funds to build capabilities at the state, local, tribal, and territorial levels to enhance overall national resilience. The SHSP supports the implementation of state homeland security strategies to address capability targets set in urban-area, state, and regional Threat and Hazard Identification and Risk Assessments (THIRAs). The capability levels are evaluated in the state preparedness report and support the planning, equipment, training, and exercise needs to prevent, protect against, mitigate, respond to, and recover from terroristic acts and natural disasters.

The purpose of the UASI program is to provide funding to address the unique planning, organization, equipment, training, and exercise needs of high-threat, high-density urban areas and assist in enhancing and sustaining their ability to prevent, protect against, mitigate, respond to, and recover from acts of terrorism.

The OPSG program is intended to enhance cooperation and coordination among local, regional, tribal, territorial, state, and federal law enforcement agencies in joint programs to secure U.S. borders along routes of access from international borders, to include travel corridors in states bordering Canada and Mexico and states with international water borders.

In addition to the HSGP grant program, the Emergency Management Preparedness Grant (EMPG) program is used to assist state and local government agencies in enhancing and sustaining all-hazards emergency management capabilities.

Unfortunately, DHS grant programs have lacked specificity when it comes to funding regional collaboration. As a result, the HSGP and EMPG programs need to provide greater clarity involving grants focusing on regional collaboration. This will help ensure that the national priority of expanding regional collaboration is a specific requirement for future grants.

The view of emergency management is that all disasters happen at the local level and that emergency response is a collaborative effort among various government agencies. Experience from recent disasters has shown that emergency management works most effectively with localized responses and solutions. Broadening that base oftentimes renders regional responses and recovery less effective, as was demonstrated in the aftermath of Hurricane Katrina.

President Bush signed a $10.5 billion relief package within 4 days of the hurricane and ordered over 7,000 active duty troops to assist with relief efforts (Committee on Homeland Security and Governmental Affairs, 2006). However, many at the local, regional, state, and federal levels charged that the relief efforts were slow because most of the affected areas were poor. There was also concern that many National Guard units were short staffed in surrounding states because some units were deployed overseas.

Due to the slow response to the hurricane, New Orleans's top emergency management official called the effort a "national disgrace" and questioned when reinforcements would actually reach the increasingly desperate city. New Orleans's emergency operations chief Terry Ebbert blamed the inadequate response on FEMA. "This is not a FEMA operation. I haven't seen a single FEMA guy," he said. "FEMA has been here three days, yet there is no command and control. We can send massive amounts of aid to tsunami victims, but we can't bail out the city of New Orleans" (Staff Writer, 2005).

A report by the Appleseed Foundation, a public policy network, found that local entities (nonprofit and local government agencies) were far more flexible and responsive than the federal government or national organizations. "The federal response was often constrained by lack of legal authority or by ill-suited eligibility and application requirements. In many instances, federal staff and national organizations did not seem to have the flexibility, training and resources to meet demands on the ground" (Singer and Howell, 2005).

Local government agencies, including police, fire, and emergency management services, have the primary responsibility for first response. As a result, it is critical that coordination at this level become the foundation for establishing corresponding regional planning and response capabilities.

History has demonstrated that regional agencies have difficulty developing wide-ranging regional planning because response activities are typically decentralized at the local level. An example of large-scale planning and response generated from the base of local responses was the evacuation of parts of New York and New Jersey during Superstorm Sandy in October 2012. Evacuation plans were based on staggered deployment of local community evacuation. Local communities had to coordinate evacuation activities to ensure a staggered evacuation during the advances of the storm. The staggered evacuation led to an effective evacuation of the many residents along the coast. The lessons learned highlight the importance for both local planning and regional coordination.

In 2003, FEMA was incorporated into DHS following the events of 9/11. From that point, FEMA's mission of disaster response was overshadowed in the consolidation and reorganization of the agency. As a result, this dramatically affected FEMA's ability to promote and support its emergency management capabilities.

In 2003, HSPD 5 mandated the formation of NIMS to provide a "consistent nationwide approach for federal, state, tribal and local governments to work together to prepare for, prevent, respond to and recover from domestic incidents, regardless of cause, size or complexity." President Bush also issued HSPD 8 (Bush, 2003), entitled "National Preparedness," that same year. HSPD 8 provides the federal government with the overall control for preparedness (NPG, 2007, pp. 11–21). To implement HSPD 8 effectively, DHS created two corresponding documents. The first was the Target Capabilities List (TCL), which established priorities, targets, and measures in order to evaluate the nation's preparedness. The second document was the Homeland Security Exercise and Evaluation Program (HSEEP). The HSEEP is a training program designed to support local, regional, and state agencies in complying with HSPD 8.

For state, regional, and local governments to have compliant exercises, they must follow the TCL and adapt their emergency management systems and programs to fit within the federal standards. Unfortunately, funding for local emergency management offices is not typically a priority for most local governments. As a result, it is hard for local and state agencies to establish a regional planning program that can comply with these standards.

As witnessed by the Geospatial Information and Technology Association's (GITA) GECCo program, regional planning will only work if local agencies recognize the need for regional planning and collaborate to define roles and create procedures that focus on regional coordination and support. This concept has been successfully applied by the Florida Department of Emergency Management (FDEM) through regional planning councils (National Infrastructure Simulation and Analysis, 2011). FDEM has coordinated with the councils across the state to map and analyze evacuations for major storm events, such as hurricanes and floods. This has resulted in improving local government emergency management coordination and response through the use of common data and a common understanding of the statutory basis that mandates regional coordination.

The city of Tampa and surrounding government agencies in the greater Tampa Bay area are an ideal example where the focus of regional planning is based on local authority and capabilities. A GECCo best practice identified during the January 2009 Tampa Bay area GECCo was that regional planning and collaboration should take advantage of the authority available at the local level and operationalize regional collaboration with local agencies on a daily basis. It was also identified that the agency personnel who first respond to an event such as a disaster are the personnel who know the details of the local jurisdictions and their capabilities.

There were several major factors identified during the Tampa Bay area GECCo that contribute to successful emergency management for a regional coordination perspective:

- Local governments need to be prepared to participate in regional coordination planning and exercises.
- Regions need to understand the existing authority of the local agencies and concentrate on regional coordination.
- State agencies need to increase local authority and require regional coordination planning.
- Federal agencies need to fund more effectively those programs that promote regional collaboration for emergency response.

As the Tampa Bay area has demonstrated, building regional collaboration is not easy to accomplish. It takes time through developing and maintaining relationships within the community. Tremendous opportunity exists for communities across the United States to improve regional planning involving emergency management and response and critical infrastructure protection. Many factors, such as those mentioned above, need to come together for successful regional collaboration to happen.

One challenge facing infrastructure owners is the increasing disparity among GIS haves and have-nots. The cost of implementing geospatial technologies was relatively high in the 1980s, which limited the use of spatial technologies to organizations—frequently large governmental organizations or utilities—that were able to justify the expenditure.

As with most technologies, those costs have dropped substantially and permitted a broader audience of users to apply the technology within their organizations. Yet, some organizations have yet to embrace the value and cost savings aspects of GIS. In the United States, this failure to act is not uncommon in smaller municipalities or rural areas where the availability of a skilled workforce may also be a limiting factor.

The growing gulf between haves and have-nots was accelerated, in part, due to large federal investments in DHS UASI and similar programs that provided substantial financial support to many major urban areas. Many GIS data sets that commonly are found in major metropolitan areas, such as land parcels and building footprints, are missing in rural areas.

This, in turn, limits the effective use of GIS and the return on investment by infrastructure owners whose assets or concerns cover large geographic areas or multiple administrative or political units. Such infrastructure owners are faced with a choice. After paying to collect needed data that would otherwise be available in a metropolis, they can share it freely or limit its distribution through policy or pricing.

While the argument about who should be financially responsible for collecting what data and how the information is shared is beyond the scope of this work, the challenge merits consideration by those using geospatial systems. Likewise, the potential consequences within the emergency management sector are tremendous, as planning must always be based upon the lowest common denominator in terms of capacity.

A major electric utility with a million customers and a well-developed GIS sharing a boundary with a 10,000-customer electric power cooperative will struggle to cooperate when recovering from storm damage or similar calamities. There is no magic solution to resolving such disparity. However, if there is recognition of the issue and of the need for regional collaboration, the potential for achieving parity and substantially increasing the effectiveness and potential return on investment exists.

5.7 Leveraging GIS for Compliance

The use of geospatial data and technology can support a variety of compliance requirements, ranging from environmental to banking and finance regulatory compliance. From a critical infrastructure protection perspective, the use of GIS for pipeline integrity management and safety provides a good example of using geospatial data and technology to assist with compliance and regulatory management. In this case, GIS helps satisfy the requirements of the Pipeline and Hazardous Materials Safety Administration (PHMSA) Distribution Integrity Management Program (DIMP) (PHMSA, 2014a) by connecting pipeline data to their geographic location.

Pipeline operators face a combination of challenging compliance and performance requirements. Examples of general compliance requirements include:

- Pipeline patrols
- Stringent construction requirements
- Operations qualification training and certification
- Data retention requirements
- Corrosion control documentation
- Class location and high consequential analysis studies
- Integrity management
- Annual reporting, including pipeline locations
- Quality control and inspections
- Inspections by state and federal auditors

There are approximately 4.2 million km of oil and gas pipeline transportation systems and nearly 1 million shipments of hazardous material by land, water, and air across the United States on a daily basis (PHMSA, 2014b). While not all pipelines are regulated, natural gas pipelines are highly controlled. For example, they are subject to an area defined and classified based upon population density and operating characteristics (for example, operating pressure, line size, material grade, and usage type) due to their risk potential. In addition, liquid product pipelines are regulated based primarily upon potential environmental impacts or high-consequence areas.

Recent accidents and subsequent changes in legislation have highlighted a need to make additional improvements in pipeline safety. For example, gas transmission pipelines in the United States have experienced an average of 78 incidents per year over the last several years (PHMSA, 2013). Such repeated incidents have driven various regulatory agencies and pipeline safety offices to implement a variety of programs for pipeline integrity management. These programs are intended to improve the safety of pipeline systems by identifying and analyzing the pipeline and facility assets that carry higher safety and environmental risks.

One such program, the National Pipeline Mapping System (NPMS) (PHMSA, 2013), requires all operators of natural gas and hazardous liquid transmission pipelines, such as natural gas distribution companies, to develop, write, and implement an integrity management program for distribution systems with the following elements:

- Define threats
- Evaluate and rank risks
- Identify and implement measures to address risks

- Measure performance, monitor results, and evaluate effectiveness
- Evaluate and report results

Along with this information, operators are required to submit pipeline locations and specific data in a GIS format on an annual basis. This information is used by a variety of government agencies and the industry to support such things as pipeline planning, route optimization, high consequential analysis, emergency response, and critical infrastructure protection.

Specific to mapping data, the NPMS consists of spatial data and associated attribute data, public contact information, and metadata pertaining to the interstate and intrastate hazardous liquid trunk lines and hazardous liquid low-stress lines, as well as gas transmission pipelines, liquefied natural gas (LNG) plants, and hazardous liquid breakout tanks jurisdictional to PHMSA (PHMSA, 2013).

The minimal accuracy of spatial data in the NPMS is ±152.4 m (500 ft). The data set includes the following attributes:

- PHMSA-assigned operator identification number
- Operator name
- System name
- Subsystem name
- Diameter
- General commodities transported
- Interstate/intrastate designation
- Operating status
- Geospatial accuracy estimate

It should be noted that the NPMS does not contain information on interconnects, pump and compressor stations, valves, direction of flow, capacity, throughput, or operating pressure. This is due in part to the proprietary nature of such data, but also to the fact that such a level of detail is not required to undertake infrastructure protection efforts.

In addition to the day-to-day operations of pipeline operators, they face a variety of challenges dealing with their pipeline and facility assets to ensure reliable supply and safe operations. To overcome these challenges, many pipeline operators are turning to GIS to provide a consolidated view of their entire pipeline infrastructure to improve such things as

- Pipeline routing and route optimization
- Pipeline planning and corrosion control
- Operations and maintenance support
- Damage prevention and one-call management

- Incident tracking
- Pipeline and corridor inspection
- Risk ranking and analysis
- Public awareness and emergency response

For the pipeline operators, GIS provides the ability to locate pipeline locations accurately, with asset characteristics and topographical and demographic data to aid in planning and route optimization. This can include development of three-dimensional models of pipeline assets using the elevation data derived from DEMs. Beyond planning, optimized pipeline routes can be developed using such information as water features, soil characteristics, transportation lines, urban centers, and environmentally sensitive areas.

GIS can also be used as part of the Integrated Pipeline Management (IPM) system for planning and replacement of pipeline and facility assets as well as assessing pipeline corrosion. For example, pipeline systems can get corroded due to various reasons, such as pipeline age and reaction of the pipe in harsh environment conditions, the type of materials being transported, soil factors for buried pipelines, and pipeline coating materials.

The contribution of GIS in support of the operation and maintenance of pipeline assets can assist with identifying locations of aging pipeline assets along with helping crews find the transmission network in the field. Integration of GIS with supervisory control and data acquisition (SCADA) systems can aid in detecting the exact locations of assets in case of a pipe burst or leak based on flow or pressure changes in the pipeline.

GIS also plays a vital role in determining the impact of hazardous liquids, gas pipeline bursts, and leaks to support high consequential analysis and emergency response activities. It supports the assessment of risks by identifying the potential high-impact zones along a pipeline corridor. GIS can provide the ability to visualize a blast or impact radius of a pipeline explosion, or study the impact of a disaster or explosion with respect to population and environmental factors.

GIS can also assist with the preparation of such things as high consequential examination, evacuation planning, and environmental impacts. Operators are using GIS to identify probability and consequence rankings for each pipe segment with regard to internal or external corrosion, excavation damage, and other operating concerns.

GIS also can be used to support inline inspections. Data can be gathered using devices called *smart pigs* that are inserted in the pipe segments to locate defects along a pipeline. The data can then be collected and reported in GIS to pinpoint necessary weld locations, as well as defects, dents, gouges, corrosion, or other abnormal operating conditions.

From an emergency management perspective, GIS is used to analyze and trend excavation-related damages for incident management. The results can be loaded into the Damage Information Reporting Tool (DIRT) to support the national program sponsored by the Common Ground Alliance (CGA) (2013). In late 2003, the CGA launched DIRT, as reported at https://www.cga-dirt.com.

DIRT is a web application for the voluntary collection and reporting of underground damage information. DIRT allows users to submit damage reports, browse data submitted by the user's organization, manage company and user information, and submit feedback and questions. Since DIRT was launched, the number of records submitted has steadily increased each year. As more operators submit data, they are able to report on the state of damage prevention throughout North America.

CGA was established to help reduce underground asset damage, which threatens safety and costs billions of dollars annually. To understand better where, how, and why these damages are occurring, accurate and comprehensive data collection is required. The primary purpose in collecting underground asset damage data is to analyze data to learn why events occur and determine how actions by industry can prevent them in the future.

The goal of this work is to ensure the safety and protection of people and the infrastructure. Data collection is used to identify root causes, perform trend analysis, and help educate all stakeholders so that damages can be reduced through effective industry practices and standards.

References

Committee on Homeland Security and Governmental Affairs. (2006). *Hurricane Katrina: A Nation Still Unprepared*. Congressional Report S.Rpt. 109-322.

Common Ground Alliance. (2013). *Damage Information Reporting Tool User's Guide*. www.Cga-dirt.com.

Department of Homeland Security. (2007). *National Preparedness Guidelines*.

Department of Homeland Security. (2009). Fiscal year 2010 Homeland Security Grant Program: Guidance and application kit.

Department of Homeland Security. (2014a). Critical infrastructure sector partnerships. http://www.dhs.gov/critical-infrastructure-sector-partnerships.

Department of Homeland Security. (2014b). Resilience. http://www.dhs.gov/topic/resilience.

Department of Homeland Security. (n.d.). *Catalog of Federal Domestic Assistance: Regional Catastrophic Preparedness*. Grant Program 97.111. https://www.cfda.gov/index?s=program&mode=form&tab=core&id=d5436b05e73dd035843cb104218d35e6.

Dwyer, J., Flynn, K., and Fessenden, F. (2002). Fatal confusion: A troubled emergency response; 9/11 exposed deadly flaws in rescue plan. *New York Times*, July 7, 2002.

Federal Emergency Management Agency. (2010). Strategic Foresight Initiative: Critical infrastructure long-term trends and drivers and their implications for emergency management. http://www.fema.gov/pdf/about/programs/oppa/critical_infrastructure_paper.pdf.

McCallister, E., Grance, T., and Scarfone, K. (2010). Guide to Protecting the Confidentiality of Personally Identifiable Information (PII): *Recommendations of the National Institute of Standards and Technology*. National Institute of Standards and Technology Special Publication 800-122. National Institute of Standards and Technology.

National Fire Protection Association. (2010). Standard for the organization and deployment of fire suppression operations, emergency medical operations, and special operations to the public by career fire departments. NFPA 1710. NFPA, Quincy, MA.

National Fire Protection Association. (2013). Recommended practice for fire flow testing and marking of hydrants. NFPA 291. NFPA, Quincy, MA.

National Infrastructure Simulation and Analysis Center, Homeland Infrastructure Threat and Risk Analysis Center, National Protection and Programs Directorate, Office of Infrastructure Protection. (2011). Tampa, Florida, hurricane scenario analysis report. October draft.

Oregon State Police–Oregon Office of the State Fire Marshall. (2014). Standards of cover. http://www.oregon.gov/osp/SFM/pages/data_standardsofcover.aspx.

Pipeline and Hazardous Materials Safety Administration. (2013). National Pipeline Mapping System. www.npms.dot.gov/datastats.

Pipeline and Hazardous Materials Safety Administration. (2014a). Distribution Integrity Management Program. www.phmsa.dot.gov/dimp.

Pipeline and Hazardous Materials Safety Administration. (2014b). For the public. www.phmsa.dot.gov/public.

Singer, L., and Howell, J. (2005). *A Continuing Storm: The Ongoing Struggles of Hurricane Katrina Evacuees*. Appleseed Foundation, Washington, DC.

Staff Writer. (2005). Katrina day-by-day recap. *Palm Beach Post*, September 1.

USA Patriot Act. (2001). Uniting and Strengthening America by Providing Appropriate Tools Required to Intercept and Obstruct Terrorism (USA Patriot) Act of 2001 [Page 115 Stat. 272]. Public Law 107-56. *107th Congress Weekly Compilation of Presidential Documents*, vol. 39, no. 10, pp. 280–285.

6

Case Study: The 2012 Republican National Convention in Tampa, Florida

6.1 Background

Tampa, Florida, is the largest city and county seat in Hillsborough County. It is located on the west coast of Florida, approximately 320 km northwest of Miami and approximately 135 km southwest of Orlando (Figure 6.1).

Tampa is the third most populous city in Florida, with an estimated population in 2010 of approximately 335,708 people. For U.S. Census purposes, Tampa is part of the Tampa–St. Petersburg–Clearwater, Florida, metropolitan statistical area (MSA). This four-county MSA, which local residents refer to as the Tampa Bay area, is home to approximately 2.7 million residents, making it the 2nd largest MSA in the state and 19th largest in the nation. The Tampa Bay designated market area (DMA) is the largest media market in the state of Florida and the 12th largest media market in the United States.

Two additional demographic components affect the use of technology in the city. The first is daily commuters. The typical daily population increase is approximately 47.5% within the city limits (Longley, 2005). This equates to an increase of approximately 159,461 people daily. The second component is seasonal visitors, or snowbirds (people who migrate south for the winter). Exact numbers are difficult to determine, but estimates consistently place the population growth during the winter months at approximately 28% in this region, or about 94,000 people in Tampa. At peak population, which also includes 900,000 short-term tourists annually, the city is providing public services and technology support to more than twice the official population.

Tampa's economy is founded on a diverse base that includes tourism, agriculture, construction, finance, healthcare, government, technology, and shipping. Geography and history have combined to create a strong regional identity and numerous mechanisms for intergovernmental cooperation and regional economic growth.

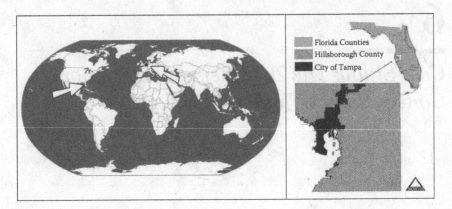

FIGURE 6.1
Location of Tampa.

6.2 GIS in the City of Tampa

The importance of geography to the city of Tampa is made clear by the city's history. The discovery of phosphate in the area in 1883 was quickly followed by the completion of the first railroad connection to the city. The city was incorporated shortly thereafter, in 1887. By 1888, railway connections were complete from New York to Tampa, promoting the expansion of the Port of Tampa and the growth of the city as a transportation hub.

As it has in many cities, the use of technology has evolved over time in Tampa. In addition, as has been common elsewhere, the sophistication of users and their demands for system enhancements have increased steadily. Perhaps nowhere else is this apparent than in the application of computer technologies to geographic information.

Tampa first recognized the significance of digital geospatial information in the 1970s. The first formal geographic information system (GIS) was developed in 1977–1978: a land use project undertaken for the Metropolitan Development Agency. Points were digitized at a local university using a Digital Equipment Corporation PDP-8 minicomputer and subsequently were input to the city's Univac 1108 mainframe computer via punch cards. These point data sets were aggregated to produce the first consistent land use map of Tampa, which was managed and displayed using the SYMAP and SYMVU digital mapping tools.

In subsequent years, that initial GIS deployment morphed into a hybrid system that became the backbone of several other citywide applications related to permitting and land use mapping. This internal development benefited from insights gained from the experience of surrounding Hillsborough County in its deployment of DeltaMap for mapping and aerial photography processing.

DeltaMap evolved from the Map Overlay and Statistical System (MOSS), which had been first developed at the USFWS in the late 1970s. The principal developer, Dr. Carl Reed, subsequently joined Autometric, where further development work produced AutoGIS and then DeltaMap. In 1989, Genasys, an Australian firm, purchased Deltasystems and the DeltaMap intellectual property from Autometric and renamed the product GenaMap. In 1994, the city of Tampa deployed GenaMap on Hewlett-Packard workstations using the HP-UX operating system at a cost of approximately $20,000 per workstation.

In 1997, Tampa transferred its core GIS data onto a MapInfo platform, which remained the primary platform for many applications until 2007. During that decade, however, decentralization of city government activities encouraged the adoption of numerous independent, department-specific platforms. Although convenient for some purposes, this approach to data management proved to be inefficient, expensive, and in some case, dangerous.

For example, the proliferation of platforms and lack of coordination in data management had resulted in the creation of 18 parallel address databases that, for the most part, were unsynchronized. The storage costs for redundant data sets were excessive, and the lack of synchronization of two databases resulted in near catastrophe on at least one occasion: a heart attack was reported at one address (a municipal park), while the paramedic's database had a different address for the same municipal park and the emergency response was delayed.

By the start of the new millennium, numerous individual departments had hired skilled GIS staff to meet their specific GIS requirements. Although a central GIS team served many municipal needs, the city had taken a predominantly vertical approach to GIS, as many of the city's departments purchased and implemented GIS technology and developed data sets independently based solely on their own needs. The result was poorly integrated applications and numerous independent databases that had little visibility among city users.

This independence was most obvious in the predominant use of two GIS platforms in multiple versions (with a much smaller number of users working on two other platforms). Data exchange between platforms required that the data be converted between the respective proprietary formats. There was no central repository for geospatial enterprise data that could store geospatial data and their associated attribute information in a format that was readily interchangeable.

This lack of a common repository resulted in redundant and overlapping data, greater long-term costs, security issues at the department level, and impaired service delivery. Integrating and controlling the quality of this GIS information was difficult due to the lack of an enterprise data model. This in turn became a major concern for the city's leadership.

With almost 25 years of experience with geospatial technology, the city was poised to move to a higher level of efficiency and effectiveness that would result in overall improvement in operations, which in turn would cascade into better service for the city's residents, visitors, and business partners. In 2002, an external consultant described the state of development:

> The City of Tampa has successfully developed an initial Geographic Information System (GIS) environment that has provided many benefits to the City. However, the City's GIS needs have grown and cannot continue to be met effectively in the current operational set up. This situation is common for a local government GIS. Many cities and counties have found that after establishing an initial GIS capability, the current situation and users' needs must be examined and a new direction must be established to effectively meet the reorganization's GIS needs. At this point, most local governments make a shift to an enterprise GIS, and this is the direction recommended for the City of Tampa. (Space Imaging, 2002)

Based on the results of this study and the potential gains from an enterprise GIS, in December 2003, the city's chief of staff for the newly inaugurated mayor chartered an internal GIS working group to draft a strategic plan to develop an enterprise-level GIS capability. The Strategic Planning and Technology Department (2004) led this cross-functional working group, which developed the city's GIS strategic plan over a 3-month period.

The strategic plan focused on four general areas: organizational structure, data standards, systems integration, and training. For each area, the plan outlined specific information for the current state, desired future state, gaps, and goals. Additionally, the team developed a near-term tactical implementation plan to set the foundational pieces in place.

The team's assessment of the state of GIS confirmed that although the city had successfully developed an initial GIS environment that coordinated and published many critical common-use GIS layers, the GIS group as it was constituted did not have sufficient resources, authority, or formal policy to integrate GIS on an enterprise level.

Overcoming this compartmentalization would require a major shift in thinking. There was a need for more coordination and citywide integration of disparate systems to eliminate duplicate efforts, to standardize the layers used by all departments, and to allow timely and accurate GIS data exchange between departments. This was to be accomplished by establishing a new enterprise-wide focus to use GIS effectively as a tool to improve the delivery of municipal services.

This enterprise approach to GIS would provide the city with many benefits:

- Increased ability to share data and conduct projects with multiple departments and external agencies
- Reduced delays in identifying resources and risks in operations of emergency-related information/processes (for example, police and fire/rescue services)

- Higher-quality data leading to the generation of new revenue from franchise tax and other tax revenue sources
- Greatly improved ability to retain the corporate or enterprise "institutional memory"
- Elimination of duplication and overlap in GIS data creation and maintenance processes, resulting in cost savings and disambiguation of conflicting data
- Application integration throughout the city
- Improved data integrity and standardization
- Faster preparation time for all geographically related displays
- Better and more easily accessed information for management decision making
- Reduced department-level expenditures to develop and administer individual data protection, security, and access/retrieval methods

The first step in establishing this focus was to define a firm authority for a centralized GIS team. This was accomplished with a city charter developed by the working group and issued by the chief of staff in 2004 (Charter, 2004). This policy reflected the city's recognition that the implementation of an enterprise GIS would require time and careful planning. The implementation had a long-term focus, but provided a short-term tactical plan to build the foundation of an enterprise-level GIS capability. The first step in the plan consisted of two parallel activities: consolidation of existing software contracts and the creation of a pilot data repository.

Regarding contract consolidation, the departmental nature of procurement had resulted in fragmentation: 21 contracts were in place to acquire and maintain four distinct GIS environments. This resulted in a loss of appropriate volume discounts in pricing, loss of discounts for training, excessive overhead costs (for example, legal review) related to contract renewals, and an excessive burden on staff resources for configuration management and user support.

The city's objective was to integrate the current computer-assisted drafting and design (CADD) and GIS platforms through the use of more advanced CADD, GIS, and database techniques to help realize one of the mayor's strategic goals of an efficient city government focused on customer service. This consolidation proceeded relatively quickly, resulting in recurring annual savings of approximately $100,000.

In parallel with contract consolidation, the city of Tampa undertook a program of consolidation for information technology staff. As was the case in many large organizations, IT in Tampa had undergone a period of decentralization during the 1990s. In 2007, with a mandate from the mayor, the newly created Department of Technology and Innovation launched a program designed to reintegrate developers from all departments and IT disciplines, including GIS, into a centralized IT production and support team. After completion in 2010, this consolidation resulted in a reduction in operating costs of more than $2 million annually.

As noted, the first GIS deployment in 1978 had been driven by a specific requirement for land use planning. Similarly, the first enterprise GIS deployment was driven by a requirement for unified utility management for the water, wastewater, and storm water departments. The utility services GIS project was seen as a way to take some quick steps to initiate a proof of concept of the central GIS relational database (in effect, a pilot repository) and provide stakeholders with results as quickly as possible. If this approach proved successful, this system and approach would be expanded to the rest of the city as appropriate. The extended utility services GIS would form the basis of the city's enterprise GIS as recommended in the 2004 strategic plan.

The integrated utility services GIS was built using the city's current technology platforms for data access and a relational database management system as a data repository. The application incorporated a set of business rules that ensured that the mapping information was topologically correct and that primary attributes had been collected before records were posted to the database. These rules helped to ensure engineering data integrity while also capturing attributes so the data were GIS ready.

To ensure data integrity, it was imperative that any applications adopted by the city be user-friendly, efficient, and contain business rules that ensured data integrity. A pilot system was rolled out to the storm water and wastewater departments for evaluation. Their rapid adoption clearly demonstrated the benefits of modern GIS solutions that could be implemented by embracing open GIS standards and enhanced the overall capabilities of these utility departments.

Existing utility data sets were migrated to the relational database for review during the training sessions. Departmental staff gave input on required data model changes after the initial test period, and final updates were made to the system rules before the system went live in less than 1 year. The water department joined in the pilot midway through the initial deployment.

The results of the pilot showed the integrity of the utility data was significantly improved by moving to a database and implementing the required business rules and integrity constraints. Adoption of the new system was rapid, as users maintained a familiar work environment. Moreover, the use of a relational database to store the core data enabled even greater opportunities for integration with other city systems, which depended heavily on an SQL Server database management system (DBMS).

The same core geographic information was linked to maintenance management, financial, and other enterprise systems used by various other departments within the city via SQL. Although no specific integration with other business systems was planned during the initial deployment, the system was designed to support anticipated future integration with an enterprise content management system and a proposed enterprise resource planning system.

In 2011, several factors arose that prompted a modest change in direction for the development of the enterprise GIS and the adoption of Esri ArcGIS as the platform for the deployment. One significant factor was the goal of direct interoperability with federal agencies during preparation for and the conduct of the Republican National Convention. Another factor was the availability of a local government data model that provided easy access to numerous shared applications—Tampa's first step toward canonical modeling of GIS data.

Perhaps the greatest single factor in this change in direction was a decision to automate the construction permitting process for the city. The financial realities of the Great Recession in the United States required cities to think innovatively about ways to foster economic growth. In the course of selecting an enterprise resource planning system, Tampa purchased an affiliated permitting system that in turn used Esri technology as its basis. It became clear that the volume of users for the permitting system justified an enterprise license, paving the way for the citywide adoption and realization of an enterprise GIS.

6.3 The 2012 Republican National Convention

Thirty-four years after the initial deployment of GIS technology in the city of Tampa, the question of the value of this approach to information management had been answered resoundingly in the affirmative. In 2006, the chief of police and mayor announced a 46% reduction in violent crime during the previous 5 years. They attributed much of this reduction to aggressive new crime prevention activities that employed GIS-based crime analysis. The reduction had reach 62% by mid-year 2011.

In the area of emergency management, the city and the adjacent Hillsborough County shared GIS data sets to facilitate evacuation planning and prepare for disaster recovery activities. In related initiatives, the city had adopted federal government standards for GIS metadata, grids, and symbology to ensure interoperability with Federal Emergency Management Agency systems during emergencies and with federal law enforcement authorities during mass gatherings, such as recent Super Bowl matches, the annual Gasparilla parade, and in 2012, the Republican National Convention. The stage was set for a real test of GIS for critical infrastructure protection.

From August 27 to August 30, 2012, the Republican Party conducted its Republican National Convention (RNC) in Tampa, Florida. The RNC and the preconvention celebration held on August 26 in St. Petersburg were designated as a National Security Special Event (NSSE) by DHS. An NSSE is an event of national or international significance that DHS judges to be a potential target for terrorism or other criminal activity.

Designation of an NSSE places the U.S. Secret Service in charge of event security. However, the local and regional governments play key roles in event security for two reasons. First, the event area that is under Secret Service control is quite circumscribed geographically and typically excludes the adjacent areas where related events such as parades, demonstrations, and speeches by the public take place.

These related events, and the physical surroundings, require security and protection services by local law enforcement agencies and, in the case of terrorist or aggressive protest activities, may require the services of fire and rescue agencies and other first responders and emergency management specialists. As a result, the federal government has traditionally provided funds to underwrite the costs associated with the provision of local security services as NSSEs.

The second reason that local and regional governments play a key role is a corollary of the first: the efforts of local agencies must be coordinated with the efforts of federal agencies. Nowhere is this more apparent than in the exchange of geospatial information. Federal agencies typically have access to highly sophisticated remote sensing systems and remotely sensed data. During a designated NSSE, for example, the National Geospatial-Intelligence Agency (NGA) is authorized to provide data to local governments that otherwise are only accessible to federal defense agencies.

Similarly, massive amounts of highly granular, detailed information are collected, analyzed, and used daily by local governments. During an NSSE, it is in the local jurisdictions' best interests to make these data available to the Secret Service and other federal agencies to ensure that their actions reflect the best and most current local knowledge and local intelligence about features, such as potable water systems, the Tampa Port Authority's storage practices, and critical infrastructure. An added benefit to sharing local data with federal agencies is the ability, if necessary, for local agencies to recover those data in the event of a disaster that damages local information storage facilities.

Given these factors, coordination of local and federal resources began about 1 year before the scheduled RNC event. Twenty-three subcommittees were formed to address specific tasks (for example, logistics). Many subcommittees were supported in their efforts by working groups (for example, the Cyber Security and Resiliency Working Group). Taking a lead position for security, the Tampa Police Department (TPD) organized a massive effort to augment local personnel with staff from other municipalities and county governments. During normal operations, TPD operates with a staff of approximately 960 sworn officers. During the RNC, the staffing level rose to more than 3,000 law enforcement officers (LEOs) working in one of three teams, as described in Table 6.1.

In overview, the RNC took place at three major venues. Tropicana Field, located in St. Petersburg and home of the Tampa Bay Rays baseball team, was the site of a massive preconvention party. The Tampa Convention Center

TABLE 6.1

Staffing Levels during RNC

Crowd Management Group	Transportation Group	Secure Zone Group
1,800 total LEOs	700 total LEOs	500 total LEOs
200 bicycle patrol	450 delegate bus security officers	150 Florida National Guard
40 mounted patrol	250 traffic control officers	170 outside agencies
		180 local agencies

was the working home for both print and broadcast media, with an estimated 15,000 reporters and staff. The Tampa Bay Times Forum, home of the Tampa Bay Lightning hockey team, was the site for the convention itself, which drew 2,286 delegates, 2,125 alternate delegates, and a total of approximately 30,000 participants from more than 100 countries.

Information technology preparations for the RNC were a mammoth undertaking. The RNC's official provider of video, high-speed data, and landline voice services placed 77 km of data cabling into service at the forum and convention center. A total of 5,000 business class phone lines were installed in the forum and convention center, and the official provider added more than 300 km of single-strand fiber to its existing cable network in downtown Tampa. Another communications carrier also added capacity to its local network. To paraphrase one local pundit, if you walked from Tampa to Jacksonville while unrolling a ball of twine, you would be placing as much string on the ground as carriers placed optical fiber in the city.

The communication system upgrades resulted in a network with the capacity to move 60 billion bits of data per second. This capacity could support the transmission of 250,000 emails per second or, given their smaller size, 37.5 million "tweets" per second. To give this perspective, a subscriber to online movies would be able to download complete high-definition Blu-ray movies in 1 s.

Another way to view this capacity: over the course of the 4-day convention, a dedicated audiophile could download every song ever recorded (more than 600 million). Just in case that wasn't enough, however, two major cellular carriers also overbuilt their local networks using cell towers on wheels (COWs).

The peak demand for electricity in the forum and convention center approached 19 MW, enough to power 7,600 homes. Fifty electricians worked to wire the forum for the convention, hanging more than 90,000 kg of lighting, speakers, and cables.

The information technology needs of the city also grew substantially. The city's Department of Technology and Innovation (T&I) developed a radio communications plan for 7,500+ public safety personnel and programmed more than 3,000 radios for field staff, officer, and peripheral relocations, including 19 command and control centers. This effort included equipping and training more than 2,300 outside officers from approximately 60 different local, state, and federal agencies.

T&I also brought 11 new systems online for the RNC, with the intention of maintaining these systems for city and regional use after the event. Among the systems deployed were

- A closed-circuit television camera and wireless mesh network used to collect evidence video and intelligence
- SAFECOP, a crime-fighting tool that drew heavily from data-sharing among officers
- A radio frequency identification (RFID) credentialing and asset tracking system
- A secure voice, text, email, and video conference tool that supported multiple functions
- A social media monitoring system, termed RNC Information Update
- An online eLearning system, delivered on a software as a service (SaaS) platform, to provide required training (and certification) for law enforcement officers prior to arrival in Tampa
- A helicopter video system upgrade from analog to digital downlink
- Enhancements to the city's field intelligence and threat analysis system

6.4 Application of GIS to Event Planning

One system of particular interest to geographers and the geospatial industry in general was TIGER. TIGER, an acronym for Tampa Information and Geographical Event Resources, is a situational awareness dashboard (also known as a common operating picture [COP]) that allowed multiple data sets to be integrated and viewed simultaneously and in near real time. Built using Flex 2.5 technology, TIGER, which was based in part on the city of Charlotte's COBRA system, is comparable to the Department of Defense's Palanterra platform and to the National States Geographic Information Council's (NSGIC) Ramona platform.

Work on geospatial data sharing began in November 2011 with a gathering of interested parties from local, country, state, and federal agencies. Work was undertaken to identify best sources of data and to define the most effective methods of data sharing and data access.

From this early coordination, the interested parties were able to identify gaps in data availability or currency and to identify methods of remediation. These geospatial data were used by federal agencies for NSSE planning and by local agencies in TIGER. These data also were served via web services to contractors working on secure, confidential systems for TPD.

Approximately 165 data sets were stored and available to TIGER users at the start of the RNC. These data sets included base maps, aerial photography, and real-time calls for service for police and fire and rescue agencies in the cities of Tampa, St. Petersburg, Clearwater, and Lakeland and the counties of Hillsborough, Pinellas, and Pasco. During the event, access was granted to the state emergency response team (SERT) data sets and to several National Oceanic and Atmospheric Administration (NOAA) data sets; the total number of TIGER data sets rose to 182 by the end of the RNC.

User response was overwhelmingly positive, with more than 350 users working both locally and in remote locations accessing the system simultaneously with no degradation of performance. Intuitive and easy to learn, T&I provided local training prior to the RNC, and an online training manual and a recorded webinar for new users were added during the event. Secure access specific to the RNC was defined, including password protection and role-based access permissions that were overseen by the city's database administration team.

Operationally, TIGER featured a pizza metaphor of crust and topping. As crust, the users could select one of two street centerline sets or one of two aerial photography data sets. This redundancy was intentional and predicated on the need for alternative data sources in the event of a point of failure in storage or access.

As pizza toppings, the users could select from the following:

- Internal data—static and real-time data developed and maintained by the city of Tampa
- Other agencies—real-time data provided by external agencies, via web services or File Transfer Protocol (FTP)
- External data—static and near-real-time data provided by external agencies (for example, National Weather Service, U.S. Geological Survey [USGS], and the Florida Department of Transportation)
- Social networks—feeds from Twitter, Facebook, and other sources, including video blogs
- Imagery—oblique aerial imagery provided by NGA for this event, as well as commercial oblique aerial and street-level imagery
- Map layers, legends, and bookmarks—including predefined event zones, preferred layer combinations, dynamic legend displays, and user-specific geo-fences
- Tools—standard GIS view and analysis tools, including the ability to upload an Esri SHP file at a single user workstation

As noted, the TIGER situational awareness dashboard was a success. Originally assembled in 12 weeks, the system was subsequently stripped down, expanded, and rebuilt using the experience gained during the RNC. At the same time, the city upgraded its GIS software platform. The resulting TIGER 2.0 provided access to more than 300 data sets, including new data feeds from the solid waste department, which implemented a route optimization system and uses TIGER to schedule and monitor real-time rerouting.

6.5 Event Activities

Was this investment in security and technology for the RNC worthwhile? The city of Tampa believes that it was. The previous RNC, conducted in Minneapolis–St. Paul, Minnesota, in 2008, was plagued by riots, violence, and criminal damage to property, resulting in 818 arrests. The 2012 Tampa RNC experienced two arrests.

Multiple factors, including the imminent arrival of a tropical storm, warm weather, and the temporal proximity of the Democratic National Convention (the last of which appears to have split the efforts of protestors), combined to contribute to the success of the convention from a public safety perspective. However, there is no question that effective, coordinated planning by TPD and the Department of Technology and Innovation reduced risk for the city and improved TPD's ability to perform its mission.

Indeed, the RNC is considered locally as just one example (albeit a dramatic one) of the effective use of information technology to achieve crime reduction in Tampa. The city has enjoyed a 64% reduction in violent crime over the past 9 years—a reduction attributed in substantial measure to effective police force deployment through the use of GIS and related technologies.

6.6 After Action Report and Lessons Learned

The primary lesson learned with respect to GIS for critical infrastructure protection (CIP) was the benefit of deploying a common operating picture accessible by all event participants. The ability to share information across the city, between multiple agencies and multiple RNC operations centers in multiple cities, often meant the difference between appropriate and inappropriate actions (both preventive and responsive).

The combination of (1) a solid digital map base with scores of data layers, (2) supplemented with fixed-mount cameras, cell phone cameras, social media broadcasts, and helicopter video casts provided the means of

TABLE 6.2

State of Florida Emergency Support Functions

ESF 1: Transportation	ESF 7: Unified logistics	ESF 13: Military support
ESF 2: Communications	ESF 8: Health and medical	ESF 14: Public information
ESF 3: Public works	ESF 9: Search and rescue	ESF 15: Volunteers and donations
ESF 4: Firefighting	ESF 10: Hazmat	ESF 16: Law enforcement
ESF 5: Info and planning	ESF 11: Food and water	ESF 17: Animal services
ESF 6: Mass care	ESF 12: Energy	ESF 18: Business, industry, and economic stabilization

TABLE 6.3

National Response (FEMA) Emergency Support Functions

ESF 1: Transportation

ESF 2: Communications

ESF 3: Public works and engineering

ESF 4: Firefighting

ESF 5: Information and planning

ESF 6: Mass care, emergency assistance, temporary housing, and human services

ESF 7: Logistics

ESF 8: Public health and medical services

ESF 9: Search and rescue

ESF 10: Oil and hazardous materials

ESF 11: Agriculture and natural resources

ESF 12: Energy

ESF 13: Public safety and security

ESF 14: Long-term community recovery

ESF 15: External affairs, standard operating procedures

analyzing the situation and, (3) used by subject matter experts sharing a common work space, was a powerful analytical support mechanism. With regard to this last component, the city used the state of Florida's emergency support functions categories to organize staff at the emergency operations center (Tables 6.2 and 6.3).

The common operating picture had to overcome four fundamental *information access challenges* that characterize most efforts to integrate and present information in real time. The first challenge was *physical access to data*, which typically reside in multiple locations on multiple storage devices or mechanisms. Tampa's response was to use commercial off-the-shelf (COTS) software and a simple three-tier data access model to aggregate the data for consumption (Figure 6.2).

City staff created and customized widgets for computer-aided dispatch (CAD) feeds from several first responder agencies in the region using FTP,

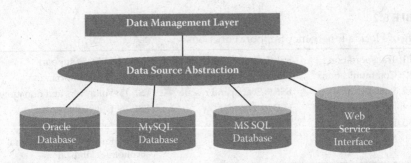

FIGURE 6.2
Data access.

web services, and direct database reads. City staff also created (1) operational layers for the Tampa–St. Petersburg area to consume RESTful services from a variety of sites; (2) launch pads for appropriate windows to view National Hurricane Center, NOAA, and National Weather Service data; and (3) widget services for real-time display of the Environmental Protection Agency (EPA) and USGS data as well as maritime vessel tracking service.

The second challenge was related to *naming conventions*, a problem related to the first challenge. As independent systems are developed over time within any organization, data requirements, characteristics of the software tools used to build the systems, or other variables often result in the use of different feature names with different formats being deployed in different data sources.

As a simple example, System A may refer to LastName, while System B uses Surname and System C uses Namefield2 (Figure 6.3). The issue was addressed during the construction of the data access model by mapping the underlying data sources into a common schema using commercial tools and web services, combined with the logic underlying the National Information Exchange Model.

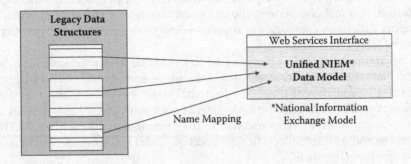

FIGURE 6.3
Name mapping.

The third challenge was that of *policies restricting access to data*. Although not a factor for all data, getting permission to share information is often harder than technical issues of sharing information. The city developed a three-part program to address this challenge:

1. Allow ample time during program implementations to obtain the appropriate access permissions. Preplanning, which may involve legal review of memoranda of understanding between jurisdictions or agencies, is critical to ensuring the availability of sensitive information.

2. Use data snapshots provided by data owners. When necessary, the city was prepared to use snapshots of data and web services to reduce concerns about intrusions in data for which other agencies were responsible.

3. Put control of what is released in the hands of data owners. A corollary of the first step, this is particularly important for public safety agencies for which specific restrictions may apply. The city's database administration staff created basic configuration roles and user profiles for the fire and rescue, police, and emergency management system staffs that strictly controlled access to the respective agencies' data.

    The staff then created a secure profile for use during the RNC. The secure profile provided access to all information to all credentialed members of the RNC operations team. This role was deactivated immediately upon the deactivation of the emergency operations center.

The fourth challenge is *usability*. A major factor in the success of this tool was its comparative simplicity, which made the tool accessible to, and usable by, not only experienced GIS professionals, but also casual users who were staffing the city's emergency operations center during the event. When information is needed, users may not have time to be well trained on system use.

In addition, given the relatively infrequent need to activate an emergency operations center for a major event, users may forget what they have learned about system operation. The city's solution to this challenge was to use a web-based interface, with simple navigation tools and extensive online training support. In this regard, the need to train people nationwide was satisfied by the creation of a recorded webinar, allowing training at the convenience of remote users.

To be sure, detailed crime analysis and other computationally intensive support tools continued to operate behind the scenes of the RNC. TIGER, the city of Tampa's common operating picture, was used to display both static and dynamic map information from multiple agencies to support real-time analysis of activities during the RNC. The ease of use of this tool was demonstrated by the fact that many city users continue to visit it even when not accessing the RNC-specific information stores.

The overall flexibility of the system was demonstrated by on-the-fly modifications to the tool to accommodate additions to support Florida SERT data feeds for storm tracking and DHS HSIP data layers. Both of these data sets were enabled after the start of the RNC as representatives of SERT and DHS began to use the TIGER system for the first time.

No system is perfect, even for a moment, and technologies continue to evolve, rendering once optimal systems in need of update and enhancement. Several areas were identified during the RNC, as summarized here.

The addition of data sets continued through the RNC and continues to the present. Nineteen data sets were added during the event; 135 were added during the subsequent 18 months, and additional data sets continue to be identified on a regular basis. Aside from the challenges discussed earlier, the volume of data available can become an issue: more data does not always translate into more information or more understanding. The identification of policies and roles for data access is critical to maintaining a manageable and actionable flow of information.

Second, many users from civilian departments gained their first familiarity with GIS technology while they worked in the emergency operations center (EOC) during the RNC. A user group was formed to define ways of extending the utility of the TIGER tool and related GIS to other departments to support additional regular business functions in those departments. Success breeds demand, which must be budgeted.

Third, the explosive growth of mobile computing devices, including laptops, netbooks, tablets, and smart phones, has changed the landscape of personal computing forever. Of these, the most significant for GIS would appear to be the tablet, although smart phones and wearable devices may be widely adopted as well. The use of common operating picture and other GIS technologies and display methods must accommodate these platforms to be widely used.

One further issue complicates the question of portable mapping tools. In many cases, the portable devices used by staff may be privately owned, raising the question of device and data security. Organizations that permit personal device use, often referred to as bring your own device (BYOD), must address the issue of mobile device management. Indeed, the risk of losing a tablet and having it fall into the possession of a terrorist during the RNC was a major factor restricting their use.

In summary, the experience of the city of Tampa with GIS and the TIGER common operating picture demonstrates the significant role such tools can perform during major event planning and management. Any entity considering such an event should consider deploying systems with comparable capabilities.

References

Charter to Develop an Enterprise Geographic Information System (GIS). (2004). January 30. http://www.tampagov.net/dept_Geographic_Information_Systems/files/Signed_TGIS_Charter.pdf.

Longley, R. (2005). Many U.S. cities see huge daily population swings. About.com. http://usgovinfo.about.com/od/censusandstatistics/a/daypop.htm (accessed August 3, 2010).

Space Imaging. (2002). *City of Tampa GIS Master Plan—Final Report*. Space Imaging, Thornton, CO, March 7.

Strategic Planning and Technology Department. (2004). City of Tampa geographic information systems (gis) strategic plan. March 31.

7

Case Study: The GECCo Project in Minneapolis and St. Paul

7.1 Background

The Twin Cities Geospatially Enabling Community Collaboration (GECCo) was held on October 27–28, 2011. It was the eighth event in a series of events that were conducted by the Geospatial Information and Technology Association (GITA) across the United States beginning in 2004. Although each GECCo event was tailored to accommodate unique community circumstances, the main theme for each was facilitation of geospatial collaboration among infrastructure owners and emergency responders in support of critical infrastructure protection and emergency preparedness and response efforts.

As a result, Department of Homeland Security (DHS) programs such as the Homeland Infrastructure Foundation-Level Database (HIFLD) Working Group, Homeland Security Infrastructure Protection (HSIP) data sets, Homeland Security Information Network (HSIN), and DHS Office of Infrastructure Protection also benefited from both the results and participation in these events. Ultimately, the GECCo program was about developing an ongoing process in a community so that utilities, government agencies, private organizations, and educational institutions with geospatial information relevant to critical infrastructure and disaster response and recovery can effectively share their data and resources during times of an emergency.

As background, the greater Minneapolis–St. Paul area is composed of 182 cities and townships settled around the Mississippi, Minnesota, and St. Croix Rivers. The area is also known as the Twin Cities for its two largest cities, Minneapolis, with the highest population, and St. Paul, the state capital. The area is part of a larger U.S. Census division named Minneapolis–St. Paul–Bloomington, Minnesota–Wisconsin. It is the country's 16th largest metropolitan area composed of 11 counties in Minnesota and 2 counties in Wisconsin with a population of 3,317,308 as of the 2010 census (Figure 7.1).

The Twin Cities GECCo included 116 participants. Group composition was well balanced with participants representing four key communities in nearly equal numbers: public service, emergency services, infrastructure owners, business, and academia. In addition, individuals from five out of the seven metro counties participated, as well as individuals from the cities of Minneapolis and St. Paul. Nearly 70 different public and private

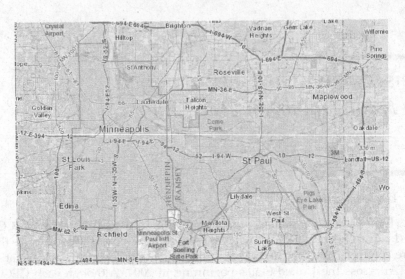

FIGURE 7.1
(See color insert.) Twin Cities metropolitan area.

organizations were represented, including utilities and other essential service providers; local, state, and federal government agencies; businesses; nonprofit organizations; academia; and community institutions.

The overall goal of the event was to raise awareness and gain knowledge of gaps in preparedness planning and management, and critical infrastructure protection in large-scale regional disasters. This was accomplished by exploring response, recovery, and restoration activities in a prolonged, cascading series of disruptions exacerbated by broad regional physical and cyber infrastructure interdependencies with major complicating factors that well exceeded the contingency planning and backup capabilities of most critical infrastructures and emergency service providers.

The Twin Cities GECCo was not designed to be an exercise in the traditional sense (that is, focused on testing existing national, state, or local plans and processes). Like similar GECCo pilots, it was not meant to follow the model that evolved from government drills that relegated private sector organizations to a lesser role, or did not include them. Rather, the GITA and GECCo regional interdependencies model relied heavily on public–private partnership and participation.

There were a number of specific exercise objectives, reflecting the diverse composition of the participating organizations. The Twin Cities GECCo was intended to

- Illuminate reconstitution and business continuity challenges and needs associated with disruptions of critical infrastructures
- Increase understanding of interdependency issues related to recovering from long-duration outages

- Underscore and validate the mutual value of public and private sector and cross-function and multidiscipline cooperation to deal with large-scale disasters

- Highlight the extent of cooperation, including understanding of roles, responsibilities, and authorities—local, county, state, and federal (civilian and defense)—of jurisdictions and private sector organizations during regional disruptions

- Increase the level of collaboration among regional emergency responders, as well as cooperation with critical infrastructure owners

- Assess what approaches and plans are necessary for improving regional data sharing and resource coordination

- Recognize and examine jurisdiction boundaries and problems that arise from these artificial barriers

- Explore the development of plans for determining restoration priorities of critical infrastructure in support of emergency response and recovery

- Identify existing laws and gaps that may impede recovery efforts

7.2 Application of GIS to Exercise Planning

The purpose of the Twin Cities GECCo was to build on the experiences and knowledge gained from previous local and regional efforts in the greater Twin Cities area to further examine and begin addressing collaboration and geospatial data-sharing issues that inhibit effective critical infrastructure protection, and emergency preparedness and response. The intent of the event was to explain and document local geospatial constraints that could affect critical infrastructure and hinder emergency responders.

In addition, it served to define how the Twin Cities geographic information systems (GIS) community can assist the emergency services and critical infrastructure protection by

- Increasing the awareness of geospatial standards and resources currently used by the emergency response GIS and remote sensing communities of practice

- Enhancing the understanding of GIS and geospatial data to support critical infrastructure protection and associated interdependencies, and emergency management

- Leveraging geospatial data into actionable information for responders and decision makers

- Identifying local initiatives and resources for improving the flow of information and geospatial data among federal, tribal, state, regional, and local data resources and stakeholders
- Gaining an understanding of the geospatial programs, tools, methods, and data available from DHS for helping infrastructure managers, first responders, emergency managers, and homeland security officials
- Examining geospatial data-sharing and collaboration issues and opportunities among public and private infrastructure owners (for example, governments, utilities, and first responders)
- Defining actionable next steps for improving collaboration, information sharing, and data quality/format needs to support more effective infrastructure protection, and emergency preparedness and response

The tabletop portion of the Twin Cities GECCo was considered the primary focus of the event because it brought together multiple elements of the learning process. Based on the following exercise objectives, the planning team focuses on the need to

- Conduct proactive planning
- Use GIS-based technologies throughout the emergency management cycle
- Employ GIS product production standards that can facilitate interoperability for the emergency services sector (ESS) and infrastructure owners

7.3 Event Activities

The Federal Geographic Data Committee's (FGDC) three key components of the National Spatial Data Infrastructure (NSDI) were used to organize the event: technology, people, and policies. The overall structure of learning included

- An introductory series of lectures providing workshop goals and objectives, lessons learned from previous GECCos, and an overview about the application of geospatial information technologies for critical infrastructure protection and emergency management planning, mitigation, response, and recovery. (This portion of the program defined the scope of work and established the basis for the remainder of the GECCo event.)

- An overview of federal efforts affiliated with the Infrastructure Information Collection Division (IICD) to help participants understand basic DHS geospatial programs and resources, and provide information on at least one federal remote sensing capability.
- A presentation provided to help participants understand the need to use standards in geospatial efforts that support the emergency services sector.
- An assortment of local stakeholder presentations and a follow-on panel discussion that outlined local/regional geospatial programs, providing networking opportunities, creating awareness about local and regional emergency response activities, and facilitating an open discussion about barriers to collaboration and data sharing.
- A tabletop exercise that encouraged discussion about interoperability needs; barriers to collaboration among local, region, state, and federal programs; and overall awareness of the geospatial needs of infrastructure owners and first responders.
- A roundtable discussion of lessons learned during the previous day and one-half of workshop activities. Facilitators gathered and integrated key points from this discussion and used them to formulate the basics of recommendations for improving collaboration.
- A summary session in which local, regional, and state-level decision makers and infrastructure and emergency management were presented with the overview of findings and recommendations.

This process used two different tracks to reach two very different core groups. Practitioners participated in all events, while decision makers participated in only the final two events.

It should be noted in assessing the results of the Twin Cities GECCo that a few of the shortfalls identified in this exercise were highlighted by other regional exercises. The exercise scenario focused on a catastrophic disaster involving multiple tornados cutting through the metropolitan area. This approach was used to fulfill the requirement to explore, identify, and assess what needed to be done to make the region as resilient as possible to a major natural disaster.

Similar to other GECCos, participants were interested in examining data-sharing and collaboration needs and challenges after a major regional disaster in a situation where there was extensive damage and disruptions of critical infrastructures and other essential services, including a prolonged power outage. To be as useful as possible, the exercise was planned to cover the response, recovery, and restoration phases of the disaster. To be as comprehensive as possible, the scope of the exercise included what prevention and mitigation measures already were in place in the region and their effectiveness in addressing a major scenario.

The participating organizations were assisted significantly by having access to nationally and locally known disaster planning, emergency response, and critical infrastructure protection experts and subject matter experts in the development and use of GIS data and technology. The participating organizations also benefited from having the experience of developing the previous regional exercises and particularly from the trusted relationships that evolved in collaborating regularly on regional infrastructure security and disaster preparedness issues.

The Twin Cities metropolitan region catastrophic wind (no. 10 major hurricane—modified) tabletop exercise "Mayday, Mayday" was developed to test the Twin Cities metropolitan region's planning, communications, and information-sharing and dissemination capabilities. The planning team decided to develop a tabletop exercise that would encourage development of public–private relationships, facilitate preventive planning discussions, and simulate a quickly unfolding set of events from a disaster. For the first GIS-based tabletop conducted in the Twin Cities, a multiple tornado scenario was selected for its applicability to climatic conditions in the region.

The tabletop exercise objectives were created to highlight awareness to

- Conduct preemptive planning
- Leverage geospatial technologies throughout the emergency management cycle
- Employ geospatial data and product production standards that will facilitate interoperability for the ESS and infrastructure owners

7.3.1 Tabletop Exercise Overview

Participants were divided into four work groups and given a developing weather scenario the evening before the tabletop to assist with advanced planning within their respective areas of interest and experience. The exercise began the following morning after participants received a review of available GIS-based technologies that could be of value to them throughout the emergency management cycle, including preparedness, response, and recovery. Specific consideration was given to issues related to preemptive planning within a multijurisdictional event, information sharing, interdependencies of infrastructure, and collaboration.

The results of the Twin Cities GECCo provided tremendous value based on the event goals:

- Networking among peers, across multiple sectors
- Providing education and greater awareness of how GIS can be used for supporting emergency response and protecting critical infrastructure

- Defining actionable goals—to improve the use of geospatial information in the Twin Cities
- Facilitating an environment that institutionalizes geospatial data/ technology/practices over time

In addition, a number of potential follow-on exercise efforts were identified:

- Conducting a GECCo sequel event focusing on hands-on practical use of critical infrastructure and key resources (CIKR)–related geospatial systems such as HSIN, HSIP, OneView, DHS Earth, and U.S. National Grid (USNG)
- Using locally managed geospatial-focused tabletops as ongoing learning and evaluation events
- Performing an event devoted to relevant nontraditional collaborative tools such as social media, open-source mapping programs, and emergency management software

The next section provides a summary of the event results, including identifying strengths to be maintained, and builds upon and pinpoints potential areas for further improvement and support development of corrective actions.

7.4 After Action Report and Lessons Learned

The following summarizes of the results from the final report of the performance of exercised capabilities and activities (GITA, 2012). The objectives of the Twin Cities GECCo event link directly to the capabilities listed below. Each capability has associated activities, corresponding observations, and recommendations.

7.4.1 Capability 1: Planning

Capability summary: As defined by the DHS Target Capabilities List:

Planning is the mechanism through which federal, state, local and tribal governments, non-governmental organizations (NGOs), and the private sector develop, validate, and maintain plans, policies, and procedures describing how they will prioritize, coordinate, manage, and support personnel, information, equipment, and resources to prevent, protect and mitigate against, respond to, and recover from a catastrophic event. (DHS, 2007, p. 21)

Activity 1.1: Conduct strategic planning

Observation 1.1: There was no regional implementing authority (IA) that directs geospatial technology development for emergency preparedness and response purposes.

Rapid advances in geospatial technologies have resulted in laws and policies that have not kept up with technical developments. There are GIS efforts that focus on individual aspects of emergency preparedness and response like 911 services of the Metropolitan Emergency Services Board, and various key organizations like GIS user groups, MetroGIS, and the Minnesota Geospatial Information Office Emergency Preparedness Committee (EPC). However, there was no specific entity that has a mandate to pull together disparate regional geospatial efforts in a cohesive way that will facilitate geospatial support of the region's ESS.

Recommendations:

1. MetroGIS should convene a work group to develop a plan that identifies an IA for the region. Specific to developing that plan, the following points should be considered:

   - Work group composition should be as diverse and as senior as possible, with adequate representation from the emergency services, public service, geospatial, and infrastructure/business communities.

   - The plan for the IA should include an organizational approach that supports Incident Command System (ICS) needs without creating duplication in existing administrative or data management structures.

   - To the maximum extent possible, the plan for the IA should use a structure that has potential for cross-community authority.

   - The work group final report should address the financial and logistic support needed to implement fully the designated implementing authority.

2. In an effort to keep Twin Cities area decision makers informed of rapid advancements in the geospatial world going forward, as well as progress on recommendations made herein, the MetroGIS work group identified above should develop a plan for keeping regional executive-level leaders informed of ongoing developments.

Activity 1.2: Develop and revise operational plans

Observation 1.2: There was currently no formal integration plan that addresses how geospatial technologies fit into local and regional command and control structures/organizations. Similar in nature to the previous observation, there existed no comprehensive plan for incorporating geospatial technologies and associated personnel into the command and control structures and organizations of the region's ESS community. Again, this is a hallmark of an emerging technology that has been addressed effectively only by a handful of organizations, such as the National Wildfire Coordinating Group (NWCG), where partners have long been required to work together collaboratively to address mutual emergency response needs through deployment of cutting-edge approaches to geospatial technology.

Recommendations:

1. Upon designation of an IA as discussed in Activity 1.1, the IA should complete an overall needs assessment and corresponding plan to incorporate geospatial technologies into local and regional command and control structures/organizations. Specific to the IA developing a plan for this issue, there is currently a lack of dynamic, ongoing discussions between the ESS and geospatial communities and decision makers.

2. As part of the overall effort going forward by the IA, a standard operating procedure (SOP) document should be developed that addresses training and operational standards.

Activity 1.3: Validate plans

Observation 1.3: There was no effort to incorporate geospatial capabilities formally into exercises conducted on the local and regional levels. Nearly every attendee of the GECCo indicated the event was the first time they had participated in a disaster tabletop exercise with a geospatial information and technology focus. Moreover, despite many of the senior geospatial community attendees, few had ever participated in any kind of disaster exercise or seen geospatial planning brought into play during those events. Unless the geospatial and ESS communities work together to bring geographic situational awareness elements into exercises, there can be no expectation that the capability will exist during a real event.

Recommendations: As part of the overall needs assessment discussed in Activity 1.2, a plan should be offered by the IA for incorporating geospatial play and teaching points into local and regional disaster exercises.

7.4.2 Capability 2: Communications

Capability summary: As defined by the DHS Target Capabilities List, agencies must have sufficient wireless communications to meet their everyday internal and emergency communication requirements. Interoperability is the ability of public safety agencies (for example, police, fire, and EMS) and service agencies (for example, public works, transportation agencies, and hospitals) to talk within and across agencies and jurisdictions via radio and associated communications systems, exchanging voice, data, or video with one another as needed.

Activity 2.1: Develop and maintain plans, procedures, programs, and systems

Observation 2.1: There were no agreed upon legal or technical protocols for the region to facilitate the exchange and use of geospatial data in support of the emergency services sector.

Without agreement on the technical and legal parameters by which geospatial data will be shared between entities in the Twin Cities area, situational awareness interoperability is incomplete at best. Hurdles include proprietary data issues, data security concerns, disparate technical capacity, and administrative and financial restrictions on data accessibility and sharing. Because practitioners cannot address these issues, engagement from the decision/policy-making community is required before there can be any meaningful sharing of geospatial data needed to provide near-real-time situational awareness.

Recommendations:

1. MetroGIS and state data practice authorities should begin discussions to create a standard memorandum of understanding (MOU) for the sharing of geospatial data between interested public and private organizations.

2. The MetroGIS work group identified in Activity 1.1 should develop recommended technical protocols to be used in conjunction with the MOU. Specific to developing a plan for this issue, the following points should be considered:

    • Specific needs of responders, local government, utilities, and industry as identified during the IA needs assessment

- Significant financial and administrative requirements by collaboratively developing and maintaining common data sets for emergency preparedness and response
- Alignment of approaches and products with approved national and state standards

3. MetroGIS should encourage champions from the decision/policy-making community to join the process.

Activity 2.2: Alert and dispatch

Observation 2.2: There was no regional standard for communicating location information by the emergency services sector. No point examined during the GECCo produced as strong a consensus as did the proposed solution for this issue. Few regions of the nation currently have a standard for communicating location during ESS operations.

As each potential solution for this issue was considered, options were eliminated until the USNG was singled out as the geospatial best practice for addressing this issue. Because this group of nearly 70 practitioners and facilitators was believed to be the most diverse and well-placed group of individuals from the geospatial, ESS, infrastructure, and business communities ever assembled in the Twin Cities, this finding is thought to be a significant outcome of the event.

Recommendations:

1. In keeping with national and state standards that create regional interoperability, the IA identified in Activity 1.2 should develop geospatial communications standards that will include use of the USNG whenever possible.

2. The IA should develop a program of outreach and education that facilitates acceptance and understanding of national geospatial standards among the region's geospatial and ESS communities.

Activity 2.3: Provide incident command/first responder/first receiver/interoperable communications

Observation 2.3: There was no regional plan for providing on-site geospatial incident support or mutual assistance. The region had yet to develop a standard for facilitating the two-way flow of real-time geospatial information to and from a location of a disaster. However, the state of Minnesota had developed GIS capabilities for deployment to the scene of a disaster, and the MnGeo Emergency Preparedness Committee had created a near-real-time remote mapping capability for supporting a disaster response.

As a way to mitigate a catastrophic loss of any specific local government's geospatial capacity during a disaster, geospatial mutual assistance agreements could be put in place between units of government in the region with similar technical capabilities.

Taken together, the above concepts represent significant ways to improve the flow of geospatial information to and from a regional disaster site while at the same time enhancing operational redundancy.

Recommendations:

1. The IA identified in Activity 1.2 should develop geospatial communications standards to facilitate the two-way flow of real-time geospatial information to and from the disaster site.

2. The IA should develop a plan for providing either a mobile GIS platform or a remote mapping production capability to support disaster responses across the region.

3. The IA should develop a regional plan to create local geospatial community redundancy.

Activity 2.4: Provide emergency operations center communications support

Observation 2.4: There was no unified flow of real-time geospatial data that would facilitate creation of a regional common operating picture (COP). A unified regional COP, or a system that could effectively share real-time geospatial data across numerous disparate viewing platforms, was needed for the region.

Although the state of Minnesota had recently been working to create a COP to support its statewide responsibilities, to date there had been only a limited effort to create a similar approach for use on a regional basis. This would provide a big-picture understanding in any major disaster that would have the potential to cross local jurisdictional boundaries.

Recommendations:

1. The IA identified in Activity 1.2 should develop a plan that will facilitate the sharing of data to support COPs in the Twin Cities region.

2. In working partnership with the MetroGIS work group identified in Activity 1.1, the IA should develop a plan for geospatial architecture and data protocols that will support creation of COPs in the Twin Cities region.

7.4.3 Capability 3: Intelligence and Information Sharing and Dissemination

Capability summary: As defined by the DHS Target Capabilities List:

The Intelligence and Information Sharing and Dissemination capability provides the necessary tools to enable efficient prevention, protection, response and recovery activities. Intelligence/Information Sharing and Dissemination are the multi-jurisdictional, multidisciplinary exchange and dissemination of information and intelligence among the federal, state, local, and tribal layers of government, the private sector and citizens. The goal of sharing and dissemination is to facilitate the distribution of relevant, actionable, timely, and preferably declassified or unclassified information and/or intelligence that is updated frequently to the consumers who need it. (DHS, 2007, p. 69)

Activity 3.1: Develop and maintain plans, procedures, programs, and systems

Observation 3.1: The Twin Cities area has an advanced collaborative geospatial community known as MetroGIS. This organization, created by the Metropolitan Council in the mid-1990s, provides the potential framework for data sharing across the region.

Recommendations:

1. Every effort should be made to leverage this strength by using MetroGIS procedures and membership as the starting point for advancing future geospatial efforts.

2. As part of the planning envisioned in recommendations for Activities 1.1, 1.2, and 2.1, consideration should be given to using the strengths of the MetroGIS model to increase the sharing and exchange of geospatial data between the public and private sectors—particularly infrastructure related.

Activity 3.2: Incorporate all stakeholders in information flow

Observation 3.2: InfraGard is a Federal Bureau of Investigation (FBI) program that is a partnership between the FBI and the private sector. InfraGard is an association of individuals, academic institutions, state and local law enforcement agencies, and other participants dedicated to sharing information and intelligence to prevent hostile acts against the United States. InfraGard chapters are geographically linked with FBI field office territories. InfraGard works with the DHS in support of its critical infrastructure protection mission, to facilitate InfraGard's continuing role in critical infrastructure protection activities and further develop InfraGard's ability to support the FBI's investigative mission.

The goal of InfraGard is to promote ongoing dialogue and timely communication between members and the FBI. InfraGard members gain access to information that enables them to protect their assets and, in turn, give information to government that facilitates its responsibilities to prevent and address terrorism and other crimes.

The Twin Cities area has an exceptional advanced public–private collaborative community (that is, InfraGard) that is interested in facilitating the exchange of information for the public good. The mission of the Twin Cities chapter of the FBI's InfraGard program "is to enable the flow of information so that the owners and operators of infrastructure assets can better protect themselves and so that the United States government can better discharge its law enforcement and national security responsibilities" (FBI, n.d.).

Unfortunately, for many private entities and utilities, whether they are participating in the InfraGard program or not, willingness to "information share" generally does not include geospatial data. Often, geospatial data are viewed as proprietary, or of such a nature that sharing would increase the chance of inappropriate use. However, the GECCo experience in other parts of the country strongly suggests that when private sector entities come to understand the limited scope of geospatial information that is needed to support regional ESS efforts, barriers are often quickly removed.

Recommendations:

1. As part of efforts related to recommendations for Activity 1.2, the IA should identify the types of data required from private sector infrastructure owners to facilitate ESS operations. In addition, provisions for data safeguards should be established, including MOUs.

2. Based on the InfraGard model, the MetroGIS work group identified in association with Activity 1.1 should offer options for increasing the number of private sector infrastructure owners who would willingly participate in efforts to share geospatial data.

Activity 3.3: Vertical flow information

Observation 3.3: There was no regional approach that facilitated the vertical flow of geospatial data during disasters.

Events of recent years have demonstrated that local or regional disasters can quickly become matters of national importance. In response to these events, the federal government began developing HSIP geospatial data sets shortly after 9/11. As conceived, these data sets were supposed to provide ESS and related communities with *a uniform and accurate* set of base layers of geospatial information across the United States.

However, because state, regional, and local entities have different geospatial data accuracies and content, technical capacity, and willingness to "upstream/downstream" this information, the quality of HSIP data sets is generally reflective of a top-down collection effort managed by federal contractors. Furthermore, even though efforts by MetroGIS and MnGeo had facilitated the exchange of data between the Twin Cities region and state, in many cases, geospatial data flows from the local level were not been optimum. Therefore, it was important that a comprehensive regional approach that facilitates a two-way vertical flow of both static and dynamic geospatial data for ESS purposes is established.

Recommendations:

1. In conjunction with Activities 1.2, 2.1, and 2.4, the IA should develop a vertically inclusive plan for a regional distributive network of static and real-time geospatial data of value to the ESS.

2. In support of the recommendation above, the MetroGIS work group identified in Activity 1.1 should recommend a regional architecture and budget plan that will facilitate real-time geospatial data flows from ESS personnel into data services that will support COPs in the Twin Cities area as envisioned in Activity 2.4.

Activity 3.4: Horizontal flow information

Observation 3.4: There was no regional approach that facilitates the horizontal flow of geospatial data during disasters.

Although the totality of collaborative efforts described in the preceding activities clearly points to an environment where city, county, and regional geospatial and ESS communities desire to work collaboratively to create horizontal data flows, in most cases geospatial data remained "siloed." This was particularly true with regard to any real-time data that might be available. In addition, there seemed to be limited appreciation for the horizontally cascading effects that infrastructure failures can have across sectors.

Without decision-maker leadership that facilitates the horizontal exchange of endorsed geospatial data, cross-compartment sharing of data will be deterred to the detriment of the common good. As a result, the only roadblock preventing the region from implementing a plan that would facilitate the horizontal flow of geospatial information was the awareness and engagement of decision makers. And if decision makers fail to become engaged on these issues in the near term, the delta between where things were and where they need to be to effectively employ available technology to enable sharing of geospatial data during future disasters will only increase over time.

Recommendations:

1. There should be engagement by public–private decision makers on all levels to ensure recommendations as offered in the preceding activities are carried out. In that regard, ongoing engagement with organizational structures such as the MetroGIS policy board and the Twin Cities UASI planning effort are thought to be critical to success. Therefore, MetroGIS should make known the existence of this report and the issues contained herein to regional executive-level leaders through outreach efforts described in Activities 1.1 and 2.1. It is thought that decision makers will then have the necessary starting point for policy and guidance decisions to solve all preceding after action report/improvement plan (AAR/IP) issues—thereby also solving those challenges related to horizontal data sharing.

2. MetroGIS and leadership of the region's ESS and infrastructure communities should work to expand engagement with each other.

7.4.4 Conclusion

The Twin Cities GECCo proved to be a substantial learning event that can serve as a model for improving GIS-based technology, spatial data, and geospatial processes supporting the region's ESS community. Of note, the event determined the following strengths of the Twin Cities area:

- Advanced thinking and collaborative geospatial community that was open to the concepts of data sharing and process improvement (Activity 3.1)
- Progressive public–private collaborative community (specifically, InfraGard) that was interested in facilitating the exchange of information for the public good (Activity 3.2)
- Forward-thinking decision makers on many levels who were willing to champion well-defined programs that facilitate the sharing of geospatial data and services

The event also determined the following areas in need of improvement:

- There was no regional IA that directs geospatial technology development for emergency management and critical infrastructure protection purposes (Activity 1.1).
- There was no formal integration plan that addresses how geospatial technologies fit into local and regional command and control structures/organizations (Activity 1.2).
- There was little effort to incorporate geospatial capabilities formally into exercises conducted on the local and regional levels (Activity 1.3).
- There were no agreed upon legal or technical protocols for the region that facilitate the exchange and use of geospatial data in support of the ESS (Activity 2.1).
- There was no regional standard for communicating location information by the ESS (Activity 2.2).
- There was no regional plan for providing on-site geospatial incident support or mutual assistance (Activity 2.3).
- There was no unified flow of real-time geospatial data that would facilitate creation of a regional COP (Activity 2.4).
- There was no uniform regional approach that facilitates the vertical flow of geospatial information during disasters (Activity 3.3).
- There was no uniform regional approach that facilitates the horizontal flow of geospatial information during disasters (Activity 3.4).

In each case of deficiency noted above, it appeared the collective regional community had a practical solution available to it. By following through with these solutions, the Twin Cities area could become a national model for how geospatial information technologies can be employed to support the emergency response and preparedness, and critical infrastructure protection.

References

Department of Homeland Security. (2007). Target Capabilities List: A companion to the National Preparedness Guidelines. http://www.fema.gov/pdf/government/training/tcl.pdf.

Federal Bureau of Investigation. (n.d.). InfraGard program. https://www.infragard.org/ (accessed November 19, 2013).

Geospatial Information and Technology Association. (2012). GITA Twin Cities GECCo after action report/improvement plan. May 31.

8

Emergency Response and Management:
Lessons Learned on the Gulf Coast

8.1 Introduction

Emergency responders and managers must make difficult decisions on a frequent basis, especially when working in the field during a crisis. However, the latter decisions are often the easiest, as they are backed by substantial amounts of training and often experience. Longer-term strategic decisions are much more difficult to make when anticipating the future.

Decisions regarding the implementation of enterprise geographic information systems (GIS), the use of crowd sourcing, and the selection of the mobile device technology that is most suitable for field data viewing and collection are complex. Moreover, they frequently fall outside the comfort zone of even the most senior emergency management experts. When faced with a choice of budgeting for items such as emergency supplies, personal protective equipment for responders, fuel for vehicles, or GIS-based technologies, the last option often loses out.

Choices such as these must be publicly defensible. When a response fails to live up to expectations, an emergency manager or infrastructure owner may quickly find himself in front of a congressional hearing panel or similar board of inquiry. At those venues, the responsible parties may be asked to explain why they chose to invest in technology when another boat, helicopter, firefighter, or medic could have potentially saved another life.

Similarly, they may be asked if hardened infrastructure could have reduced or eliminated the economic impacts from a utility outage. While many studies exist about the potential return on investment gained through the implementation of GIS-based technologies, none do or can factor the value of a human life into their computations.

The key to arguing successfully for the use of GIS in the emergency response and critical infrastructure protection segments resides in demonstrating how everyday operational use improves day-to-day efficiencies, which subsequently enable the technology's use as a force multiplier when the crisis occurs. The following examples underscore the importance of doing so while highlighting the potential returns on investment.

8.2 Establishing GIS for Crisis/Emergency Response and Management

The Mississippi Emergency Management Agency (MEMA) had few geospatial resources when Hurricane Katrina devastated the Mississippi Gulf Coast on August 29, 2005. In fact, with a staff of approximately 60 people, MEMA had relatively few resources at all and was co-located in the Mississippi National Guard Armory, where the state emergency operations center (SEOC) was housed in a single 10 × 12 m room. It was from that site that the state's entire response to the disaster was coordinated.

Geospatial resources were limited to one DVD set of the Homeland Security Infrastructure Protection (HSIP) Gold data set, a CD set containing U.S. Geological Survey (USGS) digital raster graphics (DRGs) for the state, and one computer workstation with ArcGIS Desktop and Hazus loaded. No MEMA personnel were trained in the use of GIS-based technologies, nor did the technology play a regular role in supporting planning or training exercises. However, as events unfolded, MEMA leadership quickly recognized the need for and benefits of spatial technologies in constructing and managing a response, regardless of how impromptu such efforts would prove.

Several fortuitous circumstances aided in assembling the resources needed during the first 2 weeks of the response phase. As with most states, MEMA has the ability to request aid from other state agencies when a disaster is declared. This permitted the official assignment and tasking of numerous geospatial professionals willing to volunteer their services to mission-critical geospatial tasks.

This effort was aided significantly through the use of a well-established professional network whereby members of the state's higher education community and various state agencies were able to communicate with one another and come together when the initial need for assistance was realized. The process by which such tasking occurs within the emergency management world is called a state mutual aid compact (SMAC).

This core group of approximately 12 to 15 individuals, working 24 h a day, was quickly overwhelmed, and a secondary call for assistance was rendered using a mechanism called an emergency management assistance compact (EMAC). EMAC, first established in 1996, is a congressionally ratified organization among the 50 states and U.S. territories whereby resources may be mobilized from beyond a state's borders. Furthering the effort significantly were the significant contributions of volunteers organized by GISCorps, a program of the Urban and Regional Information Systems Association.

Twenty laptop computers loaded with GIS software had just arrived in Jackson for use in a training course and were, with the permission of Esri, the software developer, reassigned for use in supporting MEMA. In both cases, the requests for additional help were formulated as basic job or technical descriptions that aided in finding and assigning the right people and equipment to the right jobs. However, this is not how such actions are normally undertaken within the intersection of GIS and emergency management.

For the overwhelming majority of SMAC and EMAC requests, additional resources are ordered by type. *Resource typing* is a term given to describing a specific set of capabilities within the context of emergency management. For example, a Type 3 Wildland Engine refers to a fire truck carrying a minimum complement of 3 wildland certified firefighters, 1,900 L of water, 300 m of attack hose line, and the ability to pump at least 9.5 L/s at a pressure of 1,700 kPa while moving. Lower numbers generally refer to greater capacity within the resource typing system.

Typed resources are designed to be activated and deactivated in a modular fashion and "plugged in" to a known portion of the Incident Command System (ICS). Unfortunately, geospatial resources were not and, as of this writing, still are not typed within the emergency management resource system beyond their role in an incident management team (IMT).

The implication for infrastructure owners, particularly those operating large private assets and facilities with their own initial response resources, such as those found at a refinery or major electric generating facility, is that, as mentioned previously, all incidents are managed locally. Directors of safety teams or private emergency responder groups must be familiar with not only the Incident Command System, but also the means and mechanisms by which additional help may be requested. Geospatial support is often overlooked when requesting assistance, especially if incident management is performed through locally built agreements and arrangements and an IMT is not requested as part of assistance plans.

In the case of Hurricane Katrina, the geospatial support team employed grew to incorporate approximately 75 individuals (Figure 8.1), some of whom occupied space in the SEOC (Figure 8.2), in a mobile computing lab housed in a recreational vehicle (affectionately called the Brain Bus) (Figure 8.3). Other members of the team worked at one of several off-site facilities managed by the state's higher education institutions or located in the six coastal counties most affected (Jackson, Harrison, Hancock, Pearl River, Stone, and Greene Counties).

The equipment needed to undertake the response was donated for temporary use by Esri, local businesses, and state and local government. Although successful in this instance, this "improvise and adapt" approach is less than optimal, and requests for GIS support during a disaster should include not only personnel and their requisite skills, required equipment, and software, but also the supplies needed to keep such help alive and fit for a period of not less than 3 days.

Geospatial support staff employed during a crisis must ensure they deploy with adequate food, water, and shelter. However, coordination with planners should establish primary and secondary provisioning schedules

FIGURE 8.1
(See color insert.) GIS group at MEMA during Katrina.

FIGURE 8.2
MEMA SEOC during Katrina.

FIGURE 8.3
Brain Bus.

for maintaining appropriate stocks of paper, ink, and similar tools of the geospatial trade. To do otherwise renders the geospatial professional in the field ineffective while increasing the food, water, and shelter supply burden placed upon emergency managers.

Geospatial professionals supporting critical infrastructure and emergency management needs during Katrina and subsequent disasters in Mississippi learned that the roles and responsibilities of geospatial support can be defined using plain language and in a manner consistent with ICS. Whether managing critical infrastructure or supporting emergency responders, a geospatial support group must fulfill the following to provide adequate 24/7 support:

1. Geospatial management and administrative support team.
   a. A team manager and one or more deputy managers are needed to provide continuity and consistency in leadership. One of these individuals should be on duty each 12-h shift. They are responsible for authorizing and assigning tasks to the team; coordinating staffing schedules; ensuring the safety of all team members; integrating workflows and products into the overall plan, protective action, or response; reporting on activities; ensuring adequate staffing, supplies, and facilities; and all other management tasks as needed.
   b. A logistics and finance specialist must requisition appropriate food, water, shelter, and supplies. This can often require budgeting, filing reams of paper or electronic requisitions, and interacting with the Emergency Management Agency (EMA) logistics

and finance staff. The logistics staff is also responsible for the physical delivery of hard-copy products when needed.

c. One or more administrative support staff members should establish a geospatial services desk. The geospatial services desk is the outward-facing single point of contact for all support requests. This role encompasses preparing or identifying a process whereby geospatial support can be centrally requested and a time estimate is quickly provided for return products or services. Once a task is undertaken, the service desk must track time on task and ensure product or service delivery.

2. Data development and management team.

a. A data coordinator is required to manage source streams, ensure access to existing collections (for example, HSIP), confirm proper metadata construction, and implement and manage an appropriate security schema whereby both data access and data integrity are ensured. The data coordinator must maintain an up-to-date listing of holdings and sources and must coordinate the delivery or availability of data for the production team. As with all other leadership positions, the data coordinator is responsible for the health and well-being of the team.

b. One or more source specialists are required to identify and obtain access to potential data sources. This task often encompasses polling other teams within the EMA or infrastructure management organization to solicit data, physically going to the field to collect data, monitoring data and news feeds/list servers, and coordinating with other external geospatial professionals. The source specialist is also charged with fielding data requests, as coordinated by the geospatial services desk, and identifying and securing appropriate sources. This may require extensive knowledge of different external data management systems and sensor systems (for example, the international charter and the satellite systems that service it). Data will not magically appear for geospatial specialists to use—they must be actively sought out.

c. A spatial data administrator must ensure the availability, versioning, and integrity (that is, consistent backup) of all spatial data. Furthermore, he or she must ensure an adequate means of data delivery to the production team. The diverse nature of spatial data may require the assignment of more than one person because this role encompasses a broad range of tasks. These include managing raster data catalogues and mosaics, establishing and maintaining data portals and services, creating and managing enterprise GIS spatial databases, and performing other advanced tasks that may require technical specialization.

3. Production team.

   a. A production team leader is responsible for coordinating with other team leads to ensure requests from the GIS service desk are satisfied in a safe and timely manner. The production team leader must also assist with tracking time and material expenditures and coordinating data requests. As different elements of the production team may work in different physical locations, the production team leader must work to ensure that adequate staffing coverage, equipment/software availability, and transportation and housing requirements are conveyed to the administrative team and met. Last, but not least, the production team leader must understand the capabilities and abilities of his or her team; use the team's knowledge, skills, and abilities appropriately and efficiently; and maintain adequate quality control.

   b. The production team may be called upon to perform a wide variety of tasks in either an infrastructure protection role or emergency response/management capacity. This includes aerial photo interpretation, photogrammetric measurement, geovisualization, use of LIDAR and other remotely sensed data, creation of analytic products, situation reporting, modeling, scripting, creation of web maps and services, production mapping, and a host of other technical tasks. Personnel may be called upon to fulfill these roles in an office environment, in the chaos of an operations center, or in the less than ideal conditions of the field.

   c. A quality assurance team must be designated. One or more individuals should be assigned the role of performing quality assurance. This encompasses basic tasks such as checking that adequate marginalia are included with printed map products (for example, U.S. National Grid [USNG], scale bar, authorship, production date, data sources, and named area[s] of interest). It also requires checking that standards are followed where required (for example, symbology, USNG overlay, page size, and scale), and ensuring that simple mistakes that might result in an unintended injury or loss are caught before any map product or service is released.

   d. A template and production mapping element are often needed to maximize team efficiency. Smart use of templates and automated procedures reduces the risk of error while speeding production time. Integrating such procedures with the production effort may be highly beneficial, particularly within enterprise GIS that are used for infrastructure management and protection or during large disasters or events.

4. A technical support team must be identified and supplied with appropriate hardware and software. The technical support team is the glue that binds the entire geospatial support group together. It should be comprised of individuals who are able to troubleshoot problems related to geospatial technologies and all of the supporting technologies required to keep them running.

   The responsibilities of the support team include software license management, reconciliation of software versions and platforms, and the provision of a hardware infrastructure. They must also coordinate with other support staff who may be responsible for ensuring Internet access, water services, electric power, and adequate heating/cooling, all while communicating needs to the administrative support team.

While GIS-based technologies are normally a relatively safe occupation choice, they become a bit riskier when employed for infrastructure management or emergency management. Even under optimal conditions for infrastructure management, field data collection may involve working in or around open trenches, heavy machinery, or volatile substances. Other issues may include enraged wildlife (for example, bees whose hives have been torn apart by high winds) and distraught homeowners with chain saws attempting to remediate a disaster that is still in progress.

When combined with an emergent situation, field hazards are no different than those faced by trained first responders. Of the many recommendations and best practices presented in this work, we cannot emphasize strongly enough the need for continual, high-quality, stringent safety training for geospatial professionals working in the critical infrastructure or emergency response and management fields.

It should also be noted that the outline above applies to roles and not to staffing levels. During normal daily operations, all of the responsibilities described may be performed by one person. During a disaster, many individuals may be required to fill a single role. This is the fundamental nature of emergency management and a duality that must be considered when planning for a GIS. Under more optimal circumstances, planning for geospatial support services should follow a more lengthy and formalized process.

8.3 U.S. National Grid and Symbology

Geospatial data portals, despite their many advantages and shortcomings, do well to serve as points of distribution for spatial data. Further, much focus is placed upon spatial data itself—what geographic features are depicted and their related features. Often overlooked and omitted are two fundamental attributes that, when incorporated with a data set, go a very long way toward improving usability by emergency responders: a standardized attribute that serves as a key for common symbology and a standardized attribute that clearly denotes a U.S. National Grid (USNG) coordinate pair.

Disasters commonly scour an environment, removing from it the recognizable features by which navigation is typically accomplished by responders. Street signs, house numbers, and any number of landmarks may be rendered unrecognizable or removed altogether by events such as wildfire, flood, earthquake, tornado, tropical cyclone, or act of terror. They may be unreadable or completely obscured by blizzard, ice storm, dense fog, or even persistent heavy rains at night. Likewise, many emergencies or disasters lack the courtesy of occurring at a well-defined street address or landmark. Coordinate pairs serve as the common language of location whereby a feature's location may be described for any purpose.

A critical lesson learned from Hurricane Andrew, and later explicitly reinforced in after action reports about the terrorist attacks on 9/11, the Space Shuttle *Columbia* disaster, Hurricane Katrina, and numerous other events was the need for a common operating coordinate and grid system for emergency responders.

This requirement is not unique to the emergency response community. In fact, the military learned the same set of lessons after friendly fire began to take a serious toll on Allied Forces during World War II, when each nation would use its own coordinate reference grid. There was a significant risk of confusion when U.S. forces called for supporting artillery fire from a British or other ally; in several cases, this confusion caused tragic results. In response, the U.S. Army Corps of Engineers developed the Military Grid Reference System (MGRS).

The MGRS is based on the Universal Transverse Mercator (UTM) coordinate system between 80°S and 84°N latitudes. The MGRS is a two-dimensional grid that uniquely identifies a square meter anywhere on the Earth's surface. Each MGRS zone is subdivided into 100,000 × 100,000 m sections aptly named *100,000 meter square*. Location within each 100,000 m square is provided by an even string of numbers, the dimensions of which depict the precision of the underlying measure. The first half of the number string depicts the easterly component of the coordinate measure, and the second half depicts the northerly component.

The MGRS uses the World Geodetic System, 1984 revision (WGS-84), as its reference datum. Its civilian equivalent, the USNG, uses the North American Datum 1983 (NAD-83) as its reference datum. For all practical purposes (other than surveying), MGRS (WGS-84) and USNG are functionally equivalent.

USNG is the recognized coordinate standard implemented by the Federal Geographic Data Committee (FGDC) (STD-011-2001) for use by all civilian federal entities. It has gained widespread acceptance among the emergency response community and was endorsed for use as the primary coordinate

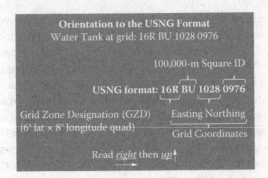

FIGURE 8.4
USNG format.

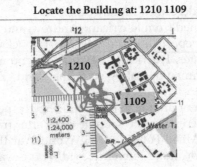

FIGURE 8.5
Reading USNG coordinates.

system for land search and rescue and air–ground coordination among responders by the National Search and Rescue Committee. Figure 8.4 depicts the format used to display a fully qualified eight-digit USNG coordinate (with 10 m precision). Figure 8.5 shows how USNG coordinates may be used to measure location.

By adding USNG coordinates to spatial data as an attribute, features may be labeled by both place name and coordinate location. Thus, when trying to communicate where something may be found in the real world or should be placed on a map, the user is empowered to do so through a standardized, widely recognizable approach. Further, use of USNG within digital systems provides a more compact means of storing or sharing location than three-attribute systems such as latitude/longitude/datum or state plane coordinate systems.

Designers of spatial data and portal systems for data dissemination must also consider attribution for map symbology. Despite the acceptance of well-defined symbol sets for use with utilities, a more broadly defined and accepted one does not yet exist for the emergency response community of practice.

Within the Department of Defense, the Tri-Services Spatial Data Standard (TSSDS) was an important first step to standardization for the Army, Navy, and Air Force. The CADD/GIS Technology Center for Facilities, Infrastructure and Environment in Vicksburg, Mississippi, developed standards collaboratively with participation from the three armed services. This work provided substantial content to the FGDC's Utilities Data Content Standard (STD-010-2000) (FGDC, 2000). Over time, TSSDS matured and evolved into the Spatial Data Standard for Facilities, Infrastructure and Environment (SDSFIE). Although standards for symbology were developed within TSSDS and SDSFIE, those standards were not adopted by the FGDC within STD-010-2000.

Efforts to develop standard symbology in the civilian agencies were active through 2007 via the Program Partners Working Group, a joint effort of USGS, the National Geospatial-Intelligence Agency (NGA), the Department of Homeland Security (DHS), and the HSIP Symbology Working Group. Unfortunately, these efforts were ultimately unsuccessful.

The lack of a common symbology presents an opportunity for massive operational confusion as different organizations downloading and mapping like data sets from the same portal may well design very different-looking cartographic products. Given that more than 600 data layers are now associated with critical infrastructure and key resources, this challenge has grown quite complex. Nonetheless, the development of attribution within a spatial data set for use in symbolizing data is a key component to presenting and understanding underlying information contextualized by location using USNG.

8.4 Hurricane Isaac

Hurricane seasons officially run from June 1 through November 30, although storms have been recorded before and after those dates. Statistically, the peak of hurricane activity occurs September 10–11, but there is significant variation from year to year.

The 2012 Atlantic Ocean hurricane season was extremely active, with 10 named storms, 4 hurricanes, and 2 major hurricanes. Hurricane Isaac came ashore in Louisiana and Mississippi as a relatively weak Category 1 hurricane (130 km/h, 1 min sustained winds) during the last days of August 2012. Two nonprofit organizations, Faster-Than-Disaster and the Stand-By Task Force, worked with volunteer GIS professionals working at the Mississippi Emergency Management Agency's State Emergency Operations Center to roll out an Ushahidi "instance."

Ushahidi is a nonprofit, open-source software company that provides mapping and visualization tools for crowd sourcing. In this case, the instance enabled the public to share storm-related events and damage reports.

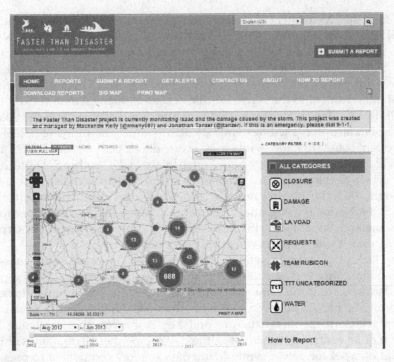

FIGURE 8.6
Ushahidi overview.

appearing in social media. While not officially sponsored by MEMA, the instance was monitored regularly by MEMA personnel and volunteers who relayed pertinent information to managers and decision makers.

An Ushahidi instance was created by Faster-Than-Disaster, and the Stand-By Task Force was used within the MEMA SEOC as a monitoring tool, as shown in Figure 8.6. The numbers inside the red circles indicate the number of reports captured for that immediate geographic area.

Figure 8.7 shows a sample report, verified by a reliable storm spotter and supported with a photograph that shows flooding due to storm surge. Such images are extremely helpful in supplementing information gained through traffic cameras and other systems.

One of the benefits of using the Ushahidi platform was that the point data collected could be sorted using basic tags and downloaded as a comma-separated values (CSV) file or Keyhole Markup Language (KML) file. The data could then be ingested by more traditional, hence widely installed, GIS platforms (for example, Esri's ArcGIS) and integrated with other data layers as needed.

One of the drawbacks of using the Ushahidi platform was that data quality was occasionally troublesome because the spatial accuracy of posts varied greatly based upon the source. Additionally, common spelling errors

FIGURE 8.7
Photo shared via Ushahidi I instance.

and mistakes with tags were present. This is a consistent issue with crowd-sourced data: while standards for data exchange are well known and used by such platforms, they are lacking with respect to qualifying positional accuracy and the quality of attribute data.

The most significant lesson learned with respect to using the Ushahidi platform was one of messaging to the public. As a first-time use of the platform, neither the public nor the media were familiar with the system and how hashtags (a word or a connected phrase with the "hash" or number sign character (#) as a prefix) could be used to mark posts on Twitter or Facebook for reporting purposes. MEMA was reluctant to include messaging about hashtags during the event itself, as the dissemination of other information, such as the location of shelters, news about mandatory evacuations, and similar time-sensitive information, was appropriately deemed a higher priority.

A secondary concern was that the public might rely on Ushahidi in lieu of E-911 or official channels for reporting damage or requesting help. The platform is not meant for such purposes, nor was MEMA staffed to do so. However, were Ushahidi to become a regular operational tool for MEMA, a messaging campaign explaining what it is and how it works would be required to maximize its benefits.

Oftentimes, where social media and public reports do not occur is as important as where they do occur. Thousands of missing person reports were filed with MEMA soon after Hurricane Katrina made landfall and prior to the complete restoration of communications services.

Reports of any missing persons, and in fact most any kind of report at all, were noted for some locations, but not all. This could be interpreted to mean that little damage and loss had occurred, that all communications systems had failed completely, or that devastation existed on a massive scale. Field teams were promptly dispatched to determine which case was true.

Social media systems such as Ushahidi can be used in a similar manner. Locations experiencing mild or moderate damage will tend to swamp Ushahidi with reports, especially where communications systems are robust. Few reports in an area surrounded by locations fielding numerous reports are often an ominous sign and warrant the immediate attention of emergency managers.

8.5 Floods Know No Bounds

While Hurricane Isaac produced a mere 2-m deep storm surge, it was a prodigious rainmaker, dropping approximately 0.3–0.45 m of rain in a 24-h period. Ground that was already saturated from earlier rains was quick to shed water and fill local waterways. SEOC personnel became especially concerned when reports of slumping and overtopping at the Percy Quinn dam were received.

A U.S. Army Blackhawk helicopter was promptly dispatched to confirm these reports and found several fire trucks from local volunteer fire departments parked on top of the spillway. The trucks were pumping water from the lake into the drainage below in a well-spirited and anxious effort to lower lake water levels and relieve pressure on the dam (Figure 8.8).

The U.S. Army Corps of Engineers immediately set about ordering much larger portable pumps to assist the fire departments. Emergency managers tasked the GIS group with modeling the potential consequences of a total

FIGURE 8.8
Percy Quinn dam failure.

failure. While Hazus was capable of performing such an analysis, the quantity and quality of data needed to produce meaningful results were not immediately on hand. Perhaps most significantly, the time required to set up and run the model was simply not available.

Initial evacuation orders were issued for occupied downstream areas. The immediate concern was that floodwaters could potentially damage bridges along the nearby Canadian National Railway and, more significantly, U.S. Interstate Highway 55, which is a major evacuation route for New Orleans.

To perform the analysis, a method similar to that used during Hurricane Katrina was quickly employed. The volume of water in the lake was estimated and, with guidance from hydraulic engineers, used to estimate the height of a flood should the dam fail (Brooks, 2005). These data were used to raise a horizontal plane in the GIS using a 10-m USGS digital elevation model (DEM) to a specified elevation above the base of the dam. The area intersected by the plane was identified as likely to flood.

Results such as these are of relatively poor quality, but acceptable for estimating potential worst-case scenarios. It is always better to use such expedient methods that err on the side of caution and then refine the results using more time-consuming but precise methods as time permits when an emergent situation is urgent.

Percy Quinn dam resides in the southern portion of Pike County, Mississippi, near the Mississippi–Louisiana state border. DEM data were not loaded into the MEMA geodatabases and had to be downloaded from Louisiana's spatial data clearinghouse. The data were then merged from quarter-quad blocks into a contiguous layer for analysis.

The results were quickly shared with emergency managers in Louisiana. Fortunately, the dam did not fail catastrophically. However, the lesson about having data on hand for areas beyond the political–administrative boundaries of immediate concern was driven home shortly thereafter. The next challenge was faced when Louisiana emergency managers notified MEMA that Pearl River lock and dam 2 was failing rapidly.

The Pearl River, which drains much of southeast Louisiana and southwest Mississippi, is a meandering and shallow waterway not suitable for navigation. To improve its use as a transportation route, a diversion canal with a series of locks and dam was constructed in the 1930s and maintained by the U.S. Army Corps of Engineers until 2005.

The waterway was turned over to local parish authorities for maintenance. The local population used the waterway as a recreational facility. Rising water soon strained the three locks located along this facility, and they too began to fail (Figure 8.9).

A complete breeching of lock and dam 2 would inundate the already strained downstream lock and dam 1 and would likely cause a massive cascading failure. Floodwaters would likely swamp the town of Pearlington, Mississippi, a small municipality that had barely recovered from Hurricane Katrina's storm surge.

FIGURE 8.9
Pearl River lock and dam failure.

As with Percy Quinn, detailed hydrologic models depicting flood depths were not readily available, nor were the cross-border data sets. The expedient methodology used to predict the effects of a Percy Quinn dam failure would not work in this instance because of the relative flatness of the area. The locks and dams are but a few meters tall, and most of Pearlington is within a few meters of sea level. Thus, an evacuation of the entire town was ordered.

8.6 Infrastructure Interdependencies: Spatial Relationships Matter

Both the Percy Quinn and Pearl River lock and dam cases provide excellent examples of the interdependencies of infrastructure from a spatial perspective. Cascading infrastructure failures are not always the result of connectivity or inter- or intrasystem relationships. The spatial arrangement of seemingly disparate systems may often have unintended consequences. Further, both cases illustrate how all disasters are local but often have far-reaching consequences.

In the instance of the Percy Quinn dam failure, a major interstate highway and transcontinental rail line (north–south) would have been immediately affected should catastrophic failure have occurred. The consequence of this local dam failure would have severely limited rail traffic between the Port of New Orleans and Canada and cut off a major evacuation and trucking route between the Gulf Coast and Memphis (a major transportation hub in its own right). Beyond wiping out a large number of homes, failure of the Pearl River lock and dam would also have affected major transportation routes and had a significant impact on agricultural production in southeast Louisiana.

A key lesson was learned from this experience. MEMA was among the very first to employ GIS to explore spatial adjacency issues in its statewide hazard mitigation plan. Using spatial data from a variety of sources, a set of topology rules were constructed to identify potential proximity issues with respect to the location of pipelines greater than 15 cm in diameter, electric transmission lines conducting more than 13 kV of electricity, and state- and federally maintained highways and railroads. This process produced an alarming number of susceptible locations. Figure 8.10 (intentionally) ambiguously, for the sake of security, depicts 6 of the nearly 100 such sites identified throughout the state.

The use of topology becomes an important tool in this regard. While some infrastructure owners currently use topology-based rules for regulatory compliance by matching jurisdictions to segments of infrastructure,

Grouped Infrastructure with Potential for Associative/Adjacency Failure

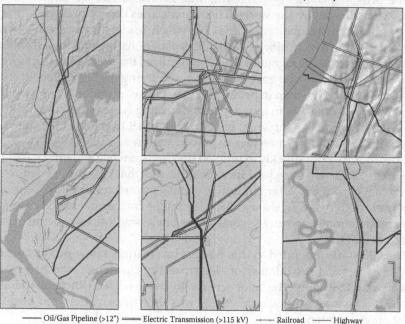

—— Oil/Gas Pipeline (>12") ═══ Electric Transmission (>115 kV) ┼──┼ Railroad —— Highway

Specific Locations Skewed for Security Purposes.
Data Courtesy Mississippi Emergency Management Agency (2013). Terrain Background Courtesy Esri.

FIGURE 8.10
Infrastructure interdependencies.

the concept may be expanded to explore similar vulnerabilities or assess potential spatially based collateral damage in the event of a failure. This consequence management approach might have exposed the vulnerabilities experienced when a 50-cm diameter gas main failed along U.S. Interstate Highway 77 near Sissonville, West Virginia, on December 12, 2012, or when a 75-cm diameter main failed and exploded underneath a residential neighborhood in San Bruno, California.

8.7 Know Your Audience

The lack of respect for political–administrative bounds extends beyond flooding. The status of electric utility service is always of keen interest to emergency managers and political leaders during a disaster. When Hurricane Katrina struck the Mississippi Gulf Coast in 2005, much of the state was plunged into darkness as electric utility transmission and distribution systems were torn apart. While at the MEMA SEOC, Governor Haley R. Barbour requested a map of power outages and the GIS team promptly went to work.

The principal challenge faced was that nearly 30 electric utilities in the state were affected by the storm. Their boundaries do not coincide with the political–administrative boundaries by which leadership wished to view the data or by which official disaster declarations are made. An intense disaggregation–reaggregation effort was required that ran well into the night.

Impatient for news, the governor began wandering the SEOC floor at about 0200 and spotted the map in progress on the GIS workstation. A red–amber–green color scheme was in use, with red depicting large numbers of customers without power. As shown in panel 1 of Figure 8.11, much of the state was without power and the map was swamped with red.

Governor Barbour quickly expressed his chagrin when showing panel 1 of this map to the news media. His reasoning was that citizens needed to see signs that even a day after landfall, progress toward restoration of services was well underway. To show otherwise might further deepen the despair felt and potentially encourage some of the lawlessness already being experienced at that time.

The GIS staff responded to this request by creating a new product for the day that depicted the number of electric utility customers who had power restored and swapped the color scheme to create a green choropleth-based product. The map, with several others, was shown soon afterward during a 0600 press conference.

The challenges associated with understanding and presenting such data truthfully are given a masterful treatment by Monmonier (1996). However, cartographers should be mindful that there are three primary audiences for GIS-based technologies services and products when disaster strikes:

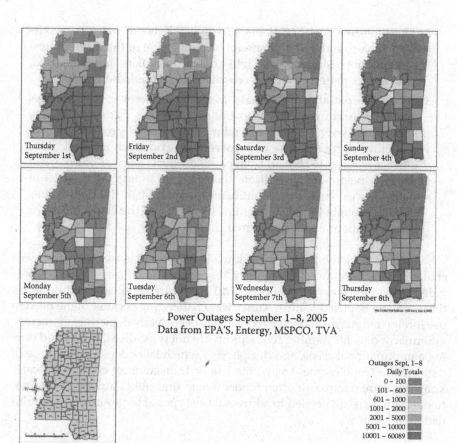

Power Outages September 1–8, 2005
Data from EPA'S, Entergy, MSPCO, TVA

Outages Sept. 1–8
Daily Totals
0 – 100
101 – 600
601 – 1000
1001 – 2000
2001 – 5000
5001 – 10000
10001 – 60089

FIGURE 8.11
Electric utility status after Katrina.

1. **Emergency managers and responders.** Maps constructed for this audience must be in a paper form factor that is easily distributed, such as A-4 (210 mm × 297 mm)-sized paper in Europe or letter (8.5 × 11 in.)-sized paper in the United States. They must use a color scheme and symbology that are clear and consistent with standards, but still comprehensible under adverse conditions such as nighttime (that is, readable in red light). They should also be drawn to a reasonable scale and contain marginalia that enable them to be used for navigational purposes or which present the coordinate-based description of areas of interest or operational boundaries. Hazards, when known, should be included as part of any base layers used.

2. **Public officials, the media, and the public.** These maps are fundamentally thematic in nature and used as *info-graphics* rather than as maps. Major landmarks and transportation features should be used to orient the reader. Where and when appropriate, the use of

coordinate overlays or callouts, such as USNG coordinates, is encouraged. Colors and symbols should be supportive of the desired message, but not disingenuous. Light, contrasting colors should be used, and intense patterns, such as crosshatching, should be avoided.

3. **Other geospatial professionals.** Maps and, moreover, map services are often used by other geospatial professionals during a crisis to select data layers, create secondary interpretive products, or for modeling purposes. Use of a coordinate grid, such as the USNG, enables features to be read from paper products (such as one marked up by a responder to indicate damaged locations). Attribution of data sources should be prominently displayed in the data frame. Standard symbols and colors must be used, and the use of offsets or representations explained in marginalia.

Languages and the use of local knowledge on crisis maps also present challenges. For example, emergency responders from many nations provided on-site assistance during the 2010 Haiti earthquake. Providing maps that correctly depicted feature names was challenging because of the need to use foreign language character sets capable of correctly depicting characters.

Similarly, data file naming conventions are not typically standardized during a multinational crisis, and deciphering which layer depicts what type of data can be troublesome. Lastly, the use or translation of common, locally known feature names can often render a map unusable. Additional international standards are needed to address these types of issues involving multinational responses.

References

Brooks, T.J. (2005). Predicting Katrina's storm surge using ArcScene. *ArcUser*, October–December 2005. http://www.esri.com/news/arcuser/1005/stormsurge.html.

Monmonier, M. (1996). *How to Lie with Maps*. 2nd ed. University of Chicago Press, Chicago.

U.S. National Grid, FGDC-STD-011-2001, Federal Geographic Data Committee, December 2001, Reston, VA.

Utilities Data Content Standard, FGDC-STD-010-2000, Federal Geographic Data Committee, June 2000, Reston, VA.

9

Use of GIS for Hazard Mitigation Planning in Mississippi

9.1 Introduction

Hazard mitigation is the assessment of risk and assignment of actions to reduce the severity of the impacts of hazard and to minimize the adverse consequences of the event. Hazard mitigation *planning* is the resultant prioritization and organization of potential mitigation actions into a comprehensive plan designed to reduce risk systematically across a jurisdiction or enterprise.

The hazard mitigation planning process, per Federal Emergency Management Agency (FEMA) guidance (FEMA, 2013), is accomplished within the following cyclic framework:

1. Determine the planning area and resources. This establishes geographic boundaries for the plan and the personnel, financial, technical, data, and analytic resources required to complete one cycle of the hazard mitigation planning process. The planning area must encompass a reasonably large (county or greater) geographic extent for assessing threats, but resultant mitigation plans and actions may only need address smaller areas, such as a utility corridor containing high-pressure gas transmission and high-voltage power lines.

2. Build a planning team with appropriate leadership. The team must include significant representation from the planning area and affiliated interests.

3. Create an outreach strategy that engages stakeholders and the public. Public meetings that engage communities in the planning area build support for the plan and provide an opportunity for constituents to provide input.

4. Review community capabilities. This extends beyond immediate first responder (which encompasses utility crews) and emergency management resources to include a total of 31 core areas (FEMA, 2007).

5. Conduct a risk assessment. This includes a formalized review of assets, threats and hazards, and vulnerabilities.

6. Develop a mitigation strategy. This may include finding individual or collective means by which assets are hardened, threats/hazards are reduced, and vulnerabilities are minimized. Strategies employed should be sustainable, and the creation and use of ordinances, building codes, or similar policies is recommended.

7. Draft the plan and keep it current. Risk is not static, and the hazard mitigation process is not brief.

8. Review and adopt the plan. Ratification and approval by the planning team and all jurisdictions in the planning area is required and provides the basis for common policy and strategies for coping with a disaster.

9. Create a safe and resilient community. Implement the projects and goals identified during the hazard mitigation planning process.

Risk comprises three components—an aggregation of assets whose loss would imperil a community, perhaps best thought of as critical infrastructure and key resources: threats, hazards, and vulnerability. Understanding risk is accomplished within the context of a Threat and Hazard Identification and Risk Assessment (THIRA), which encompasses the following:

1. Identifying potential threats and hazards of concern: What natural or man-made events may occur with potential for disaster?

2. Providing threats and hazards context: How will a threat or hazard manifest to cause a disaster within a community or system?

3. Establishing capability targets: What are reasonable and prudent means by which risk may be reduced?

4. Applying the results: What resources are required to meet capability targets?

Both the THIRA and the overall hazard mitigation planning process are inherently geographic in nature. The presentation that follows includes excerpts and examples from the MEMA Region 3 Hazard Mitigation Plan created as part of the 2014 planning cycle, which explains the THIRA process and subsequent incorporation into the determination of potential mitigation activities (Brooks and Boone, 2014).

It must be noted that the hazard mitigation process typically is where the highest return on investment in geospatial information technologies occurs within the life cycle of emergency response and critical infrastructure protection. The identification of mitigation actions using geospatial techniques and tools provides the opportunity to prevent catastrophic losses through relatively minimal investment, as will be demonstrated.

9.2 Overview of Hazard Identification

Hazard identification is recognizing risk-related events that can threaten a community. Events are described as natural or human-caused hazards that inflict harm on people or property, or interfere with commerce or human activities. Such events could include, but are not limited to, tropical storms, floods, severe storms, earthquakes, and other incidents affecting populated or built areas (DHS, 2013). A full listing of threats for the study area in Mississippi is included in Table 9.1.

FEMA requires evaluation of a standard set of hazards for its purposes. This process involves a systematic approach to the identification of natural hazards that is not likely to occur within the planning area. Additional hazards may be added at the discretion of the planning team.

Appropriate data sources for each relevant hazard should be identified. Common sources for natural disaster data for the United States include:

- Extreme heat, cold, and winter weather: National Climate Data Center (http://www.ncdc.noaa.gov/)
- Tornado, wind, and hail: Storm Prediction Center (Severe Weather GIS (SVRGIS)) (http://www.spc.noaa.gov/gis/svrgis/)
- Volcanic activity, earthquakes, wildfire, and space-based phenomena: National Oceanic and Atmospheric Administration (NOAA) National Geophysical Data Center (http://www.ngdc.noaa.gov/ngdcinfo/onlineaccess.html)
- Tropical cyclones/hurricanes: National Hurricane Center GIS archive (http://www.nhc.noaa.gov/gis/)

TABLE 9.1

Hazard Identification

Natural Hazards Not Likely to Occur within the Planning Area		
• Avalanche	• Sinkhole	• Storm surge
• Tropical storm	• Tsunami	• Volcano

Natural Hazards Prone to the Planning Area		
• Dam/levee failure	• Extreme heat	• Severe winter weather
• Drought	• Flood	• Tornado
• Earthquake	• Severe storm (thunderstorm,	• Tropical storm/hurricane
• Expansive soils	lightning, hail, and high wind)	• Wildfire

Man-Made/Health-Related Hazards		
• Hazardous material incidents	• Pandemic influenza	• West Nile

- Hazardous materials incidents and wildfire: USFA National Fire Information Reporting System (https://www.nfirs.fema.gov/)
- Flood: FEMA National Repetitive Loss Property Viewer (http://www.fema.gov/media-library/assets/documents/966)
- Hazardous materials: U.S. Coast Guard National Response Center (http://www.nrc.uscg.mil/)
- Hazardous waste identification: Environmental Protection Agency (http://www.epa.gov/waste/inforesources/pubs/training/hwid05.pdf)

Other hazards, such as the location and classification of dams or soils, are best obtained through state or local resources. Disease information may be difficult to capture and is often generalized for privacy reasons. Flood prediction mapping is best accomplished using Hazus, a free GIS-based modeling tool available from FEMA.

The Hazus tool permits users to model the potential effects of earthquakes, coastal flooding, dam/levee breach, and hurricanes (FEMA, n.d.). These natural hazard data should be addressed individually in a mitigation plan and assessed with respect to frequency or probability of future occurrence and, where available, prior loss data.

We can supplement predictive mapping based on Hazus with maps of repetitive flood loss claims that provide an initial indication of flood frequency and the probability of future occurrence (Figure 9.1). Additional analysis, such as normalization for population or structure density, may be desired.

Figure 9.2 illustrates total financial loss due to flood. As with frequency, additional analysis whereby loss type (crop/structure) may yield additional insight should be guided by the planning team.

As we learned from the effects of Hurricane Sandy, also known as Superstorm Sandy, on the New York metropolitan area in 2012, the quality of flood prediction mapping is variable. This is due in large part to the time delays that occur when incorporating updated information into existing data sets. Changes in population density change risk, as do such factors as beach erosion, channel dredging, and new building construction. Mitigation plans must be subject to regular thorough revision to ensure their suitability and effectiveness.

Mitigation planning teams should understand that certain types of threats often cover large geographic areas. Examples of such threats can include ice storms, hurricanes, and earthquakes. Conversely, hail, tornado, and severe winds often effect relatively small geographic extents, but with devastating effects. Analyzing the severity of the damage, path length and width, and bearing may provide additional insights to support mitigation planning.

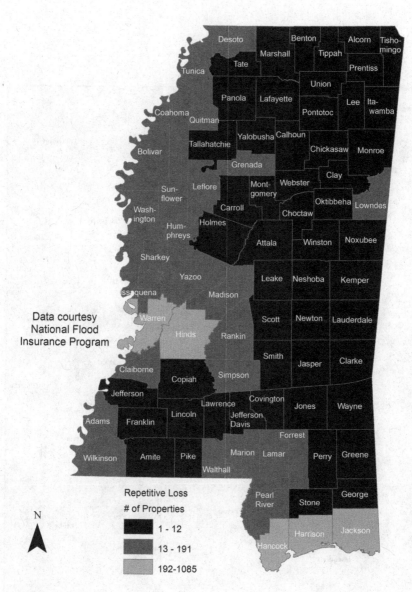

FIGURE 9.1
Repetitive flood loss claims.

Most disasters, by definition, are discrete, rare events. Tornados are relatively rare and do not happen every day in any single geographic location. Likewise, a tornado happens or it does not happen—there is no such thing as a quarter of a tornado. When compared with the bell curve, or normal distribution, the distribution of discrete rare events tends to skew to the left when plotted on a frequency histogram.

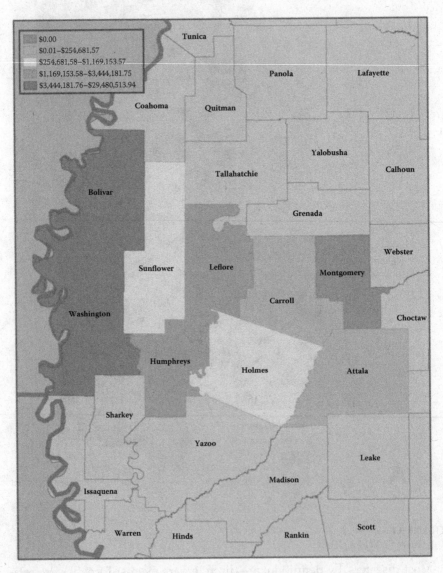

FIGURE 9.2
Total financial loss due to flood.

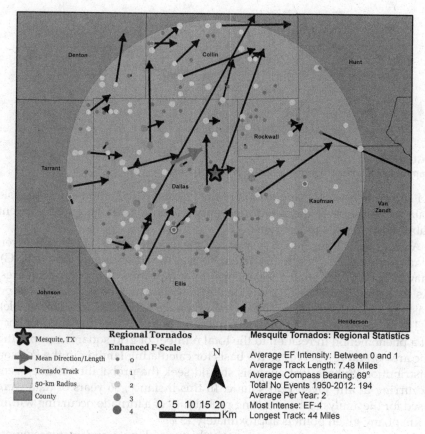

FIGURE 9.3

(See color insert.) Mesquite, Texas tornadoes (1950–2012).

This method of analysis, documented by S.D. Poisson, became known as the Poisson distribution (Poisson, 1837, p. 206). It is used to analyze the number of occurrences of an event within a given unit of time, area, volume, or similar. The Poisson distribution is defined in Equation 9.1:

$$P(x) = \frac{e^{\lambda} * \lambda^{x}}{x!} \qquad (9.1)$$

where $P(x)$ is the probability of occurrence of event frequency (x), and λ is the average number of features per unit (area, volume, time, etc.).

Figure 9.3 shows the incidence of tornadoes near Mesquite, Texas. Note the data for intensity, average track length, average bearing, and similar statistics in the marginalia and the large red arrow depicting the mean track length and travel direction in the map.

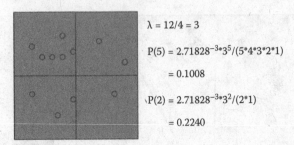

$\lambda = 12/4 = 3$

$P(5) = 2.71828^{-3}*3^5/(5*4*3*2*1)$

$\qquad = 0.1008$

$P(2) = 2.71828^{-3}*3^2/(2*1)$

$\qquad = 0.2240$

FIGURE 9.4
Application of Poisson distribution to an example data set.

Figure 9.4 shows the application of the Poisson distribution to an example data set. Assume 12 discrete rare events within a given geographic extent subdivided into 4 equal regions.

A geographic information system may be used to cast a grid for any given planning area, or a coordinate system, such as the U.S. National Grid (USNG), may be used to designate 1,000 m, or similar multiples of 10 m, grid squares. As shown in Figure 9.5, an individual grid square may be selected based on past track lengths and an assumed track damage width (500 m is reasonable).

The number of grid squares occupied by points indicating the presence of a phenomenon in relation to the total number of grid squares comprising the area of interest serves as the basis for calculating lambda in the Poisson distribution. Subsequent analysis should seek the probability of one event occurring in any given grid square. In this instance, 56 years of data were used for the analysis; the resulting probability of a tornado occurring within 1 km of any given point is approximately 15%.

An important aspect to note is that historic climate and meteorological data are often problematic in that data sets may not be spatially accurate, may contain confounding data, or may be incomplete. Consider the example of Bolivar County, Mississippi. All three problems are evidenced in Figure 9.5.

As shown, the National Weather Service determined touchdown and track position with 0.1 decimal degrees. Track data were not recorded for all events, and while the majority of tornados appear to move from southwest to northeast, a few traveled from southeast to northwest. The latter phenomenon occurs when a tornado is associated with the right-front entrance region of a land-falling hurricane, whereas the former is typically associated with frontal passages.

Close examination of Figure 9.5 also suggests a "hole" in the data near the center of the county where no tornados were recorded. Planning teams should note such phenomena, as they may bare significant consequences, as demonstrated in Figures 9.6 through 9.9. A map of population density (Figure 9.6) clearly shows low population densities in the center of the county, whereas many tornados are observed around populated places and along major roadways.

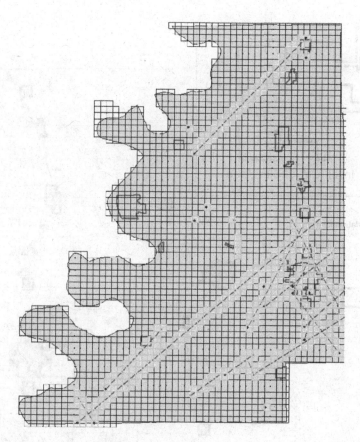

FIGURE 9.5
Tornado tracks for Bolivar County, Mississippi (1950–2006).

Further analysis determined that approximately 67% of all tornados occurring in this geographic region occur at night. An examination of effective weather radar coverage (Figure 9.7) indicates that the area of interest is at the very edge of the three closest weather radar systems. All three radar images were captured simultaneously, yet each depicts slightly different local weather conditions. Thus, it is likely that many tornado events go unobserved, especially in such largely agricultural areas where fields are bare during tornado seasons. The latter point is easily confirmed through the use of land use/land cover data derived from Landsat, an earth monitoring system managed by the U.S. Geological Survey (USGS) (Figure 9.8).

Lack of observation power is frequently a challenge nationwide when working with tornado, severe wind, and hail data sets. Figure 9.9 displays 30-km resolution aggregated tornado data for the United States, 1950–2002, as mapped by the National Weather Service Storm Prediction Center in Norman, Oklahoma, and overlaid with U.S. interstate highway information.

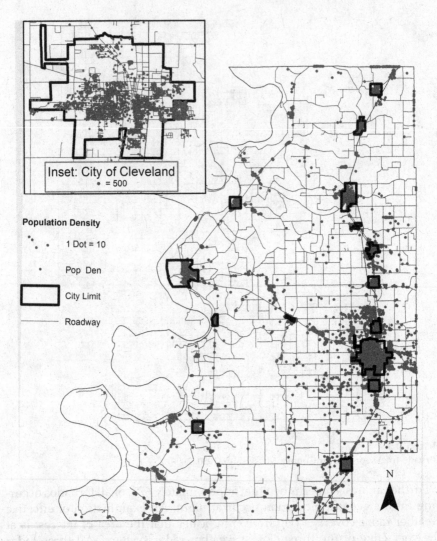

FIGURE 9.6
Population density for Bolivar County, Mississippi.

Note the tendency of tornados to follow interstate highway routes. Although clearly observable, this is likely a coincidence (if, in fact, it is a true statement). Rather, it is likely that the perceived relationship represents a historic lack of observational capabilities off road and away from major transportation corridors.

The distribution of hazard data, particularly with respect to man-made disasters such as industrial explosions, terrorist activities, and hazardous materials spills, is useful to examine from a pattern detection perspective.

NWS Little Rock   NWS Memphis   NWS Jackson

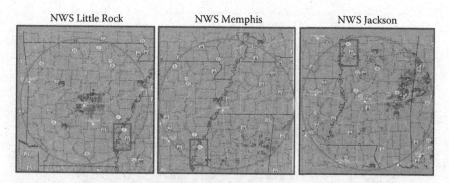

FIGURE 9.7

(See color insert.) Effective weather radar coverage for Bolivar County, Mississippi.

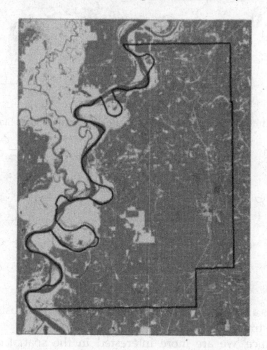

FIGURE 9.8

Land use and land cover data for Bolivar County, Mississippi.

Tobler's first law of geography simply states that near things are more closely related than distant things.

Inherent in this idea is that similar objects will tend to cluster in groups in geographic space. This principle forms the basis for many statistical tests and predictive methods, most notably Moran's I, Geary's C, and Gedis Ord Gi*. Moran's I is a test for spatial autocorrelation. It returns a value of –1.96 to 1.96.

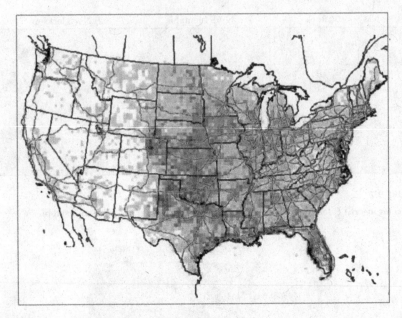

FIGURE 9.9

(See color insert.) Thirty-kilometer aggregate tornado data for the United States (1950–2002).

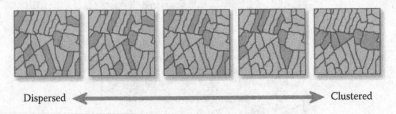

Dispersed ⟵⟶ Clustered

FIGURE 9.10
Spatial distribution.

Negative values are not spatially autocorrelated, whereas positive values are. Figure 9.10 illustrates this concept.

In this instance, we are more interested in the spatial distribution of prior disasters, asking, "Are some areas within the region more prone than others?" While Moran's I will indicate the degree to which spatial autocorrelation is occurring in association with a given phenomenon, particular interest resides in recurring clusters where future events are likely to occur, or what are more commonly known as hotspots.

The Gedis Ord Gi* statistic is useful in determining the location of hotspots and will be used both to understand the clustering of hazards, particularly those that are human caused, and as a means of detecting vulnerability (clustered key assets and critical infrastructure increases vulnerability).

Police and fire departments use this statistic frequently to determine patterns among like incident types. From a hazard mitigation perspective, analysis of human-induced disasters such as hazardous materials incidents drawn from the National Fire Information Reporting System may be particularly useful.

Similar processes and procedures should be followed for all identifiable natural hazard threats and the results presented to the planning team. It is crucial that potential weaknesses and strengths in the data underlying data or modeling processes are both well documented and shared. The presentation of a summary data table, similar to Table 9.2, is often helpful as the planning team moves toward its ultimate goal in this part of the THIRA process: the establishment of a hazard ranking and subsequent decision making about mitigation.

9.3 Providing Context: Vulnerability Assessment

Vulnerability is defined as the susceptibility of a community and its constituent social and economic groups to damage or injury from hazards (Godschalk, 1991). There are two components to vulnerability inherent to this definition: the degree to which damage may be limited and the ability of a community to recover. The latter is often referred to as resilience (Buckle, 1995). Both aspects can be analyzed from a geospatial perspective.

Regardless of approach, the identification of critical infrastructure and key resources (CIKR) within this context and from a geographic perspective is required. As with threat analysis, the identification of CIKR is an iterative process subject to ranking by the hazard mitigation planning team. No universal formula for what is important and what is not to a given community, utility system, or other element of infrastructure applies, as these features and their relative importance vary greatly.

The most direct manner to explore vulnerability is to use simple analysis: proximity to known hazard locations, such as dams or earthquake faults. Figure 9.11 demonstrates this approach. Derivative approaches may include the creation of flood depth grids through modeling exercises, as shown in Figure 9.12, which illustrates the susceptibility of major roadways and municipalities to a dam failure scenario.

It is important to note that vulnerability mapping and the messages conveyed rely heavily upon appropriate use of cartographic technique to convey meaning to the map user. As with a threat analysis, true vulnerability may become massively distorted if adequate care is not taken. This is a topic of discussion thoroughly examined by Monmonier in *Cartographies of Danger* (1998).

TABLE 9.2

Example Summary Hazard Table for the MEMA Region 3 Hazard Mitigation Plan 2004

Hazard Type	Attala	Bolivar	Carroll	Holmes	Humphreys	Leflore	Montgomery	Sunflower	Washington
Drought									
Number of events	7	5	6	6	5	5	7	5	5
Property damage	70,000	—	30,000	40,000	80,000	50,000	50,000	—	—
Crop damage	4,400,000	6,800,000	3,300,000	3,300,000	6,500,000	4,400,000	2,400,000	3,800,000	6,800,000
Death/injury	-/-	-/-	-/-	-/-	-/-	-/-	-/-	-/-	-/-
Excessive Heat/Heat									
Number of events	5	5	5	5	5	6	5	6	7
Property damage	—	—	—	—	—	—	—	—	—
Crop damage	—	—	—	—	—	—	—	—	—
Death/injury	-/-	-/-	-/-	-/-	-/-	-/-	-/-	-/-	-/-
Flood									
Number of events	12	19	9	9	10	12	14	13	23
Property damage	1,400,000	1,756,000	790,000	1,473,000	1,890,000	448,000	781,000	1,780,000	16,368,000
Crop damage	—	1,205,000	150,000	300,000	1,100,000	—	1,200,000	1,200,000	21,252,000
Death/injury	-/-	-/-	-/-	-/-	-/-	-/-	-/-	-/-	-/-

(Continued)

TABLE 9.2 (Continued)

Example Summary Hazard Table for the MEMA Region 3 Hazard Mitigation Plan 2004

Hazard Type	Attala	Bolivar	Carroll	Holmes	Humphreys	Leflore	Montgomery	Sunflower	Washington
Severe Storm									
Number of events	118	160	118	114	54	114	82	124	126
Property damage	1,708,000	4,716,000	6,556,00	533,000	576,000	6,586,000	2,020,000	9,454,000	3,969,999
Crop damage	100,000	715,000	726,500	48,000	150,000	1,350,000	200,000	3,550,000	409,000
Death/injury	–/–	–/–	–/–	–/–	–/–	–/–	–/–	–/–	–/–
Severe Winter Weather									
Number of events	10	15	9	9	8	10	7	13	12
Property damage	1,123,000	1,710,000	985,000	1,160,000	895,000	1,510,000	642,000	1,305,000	1,225,000
Crop damage	–	–	–	–	–	–	–	–	–
Death/injury	–/–	–/–	–/–	–/–	–/–	–/–	–/–	–/–	–/–
Tornado									
Number of events	12	15	5	15	9	15	5	13	7
Property damage	65,900,000	5,455,000	115,000	64,527,000	2,348,000	2,330,000	713,000	1,980,000	1,055,000
Crop damage	7,116,000	500,000	320,000	6,325,000	705,000	1,115,000	800,000	215,000	–
Death/injury	0/10	0 Deaths 73 Injuries	–/–	1/43	1/5	–/–	–/–	–/3	1/2

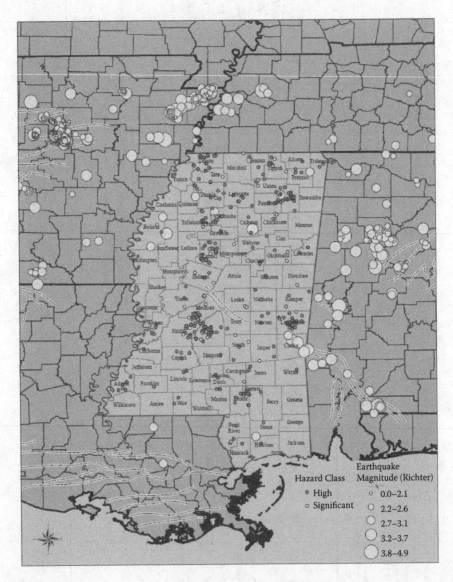

FIGURE 9.11
Regional earthquakes, normal and quaternary faults, and fault areas in proximity to high- and
significant hazard dams in Mississippi (2013).

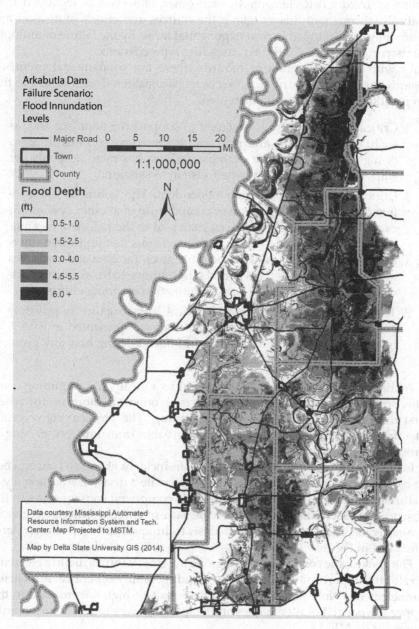

Arkabutla Dam
Failure Scenario:
Flood Innundation
Levels

—— Major Road
☐ Town
☐ County

Flood Depth
(ft)
☐ 0.5–1.0
☐ 1.5–2.5
☐ 3.0–4.0
☐ 4.5–5.5
☐ 6.0 +

0   5   10   15   20
|———————————|Mi
1:1,000,000

N

Data courtesy Mississippi Automated
Resource Information System and Tech.
Center. Map Projected to MSTM.

Map by Delta State University GIS (2014).

FIGURE 9.12
(See color insert.) Flood depth grids reflective of an Aklabutla Lake dam breech due to earthquake.

These techniques are particularly useful when determining the vulnerability of fixed CIKR elements to each other. However, as identified thematically throughout this text, both the sources and effects of disaster are commonly compounded in near exponential terms by the failure of multiple infrastructure systems and their cascading repercussions.

As with the distribution of hazards, there are fundamental premises regarding critical infrastructure that must be considered with regard to the preparation of a hazard mitigation plan:

1. Critical infrastructure is designed to support the populace and, as such, where possible and practical, is more concentrated in areas of higher population (for example, Jackson has more hospitals per square kilometer than anywhere else in Mississippi).

2. Critical infrastructure is interdependent. The failure of one system often leads to the failure or compromise of another system (for example, loss of electrical power may lead to the failure of communications, water, and sewage systems). Understanding these interdependencies is typically not possible given the data and resources available for analysis. This is, and will continue to be, a critical shortcoming of this and future hazard mitigation planning efforts.

3. The overwhelming majority of critical infrastructure is privately held. This not only compounds the challenge presented in item 2, but creates significant blind spots in understanding how any given threat will affect vulnerability.

While spatial statistics may prove useful as a means for examining complex scenarios, simple proximity analysis is often far more useful when first delving into hazard mitigation planning. The physical convergence of infrastructure elements is readily observable from imagery or simple ground reconnaissance.

Even in the absence of external events, an incident at such a convergence or coincident geometry has the potential to create a disaster. Additionally, a naturally occurring event may significantly compound consequences in the absence of appropriate risk reduction measures. Such physical convergences are easily identified using spatial analysis techniques and tools, clearly demonstrated in Figure 9.13.

The geographic convergence of significant critical infrastructure elements creates the opportunity to accelerate cascading failures of critical infrastructure elements. Note the proximity of features to a high-hazard dam in the upper-left panel. Location names and map scales were redacted for security purposes by MEMA.

Ultimately, all vulnerability studies must incorporate information about the susceptible population and its ability to recover. Initial steps should include mapping basic metrics such as population density, housing stock, day versus night population, institutionalized segments of the population

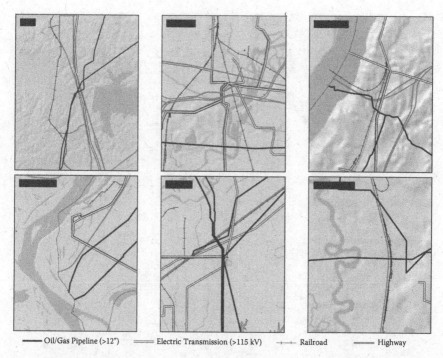

— Oil/Gas Pipeline (>12")   ═══ Electric Transmission (>115 kV)   ┼─┼ Railroad   ── Highway

FIGURE 9.13

(See color insert.) Geographic convergence of significant critical infrastructure elements.

(those in jail, a hospital, nursing home, or similar facilities), and population segments especially vulnerable to a crisis, such as the elderly or very young. Input from the mitigation planning team is particularly useful in determining vulnerabilities of concern.

As illustrated in the hazard identification section, tornados are a significant concern in Mississippi. Lightweight building construction is common in the region due to poor economic conditions. A map of mobile home housing stock, as shown in Figure 9.14, may be especially useful in understanding vulnerability to many hazard classes.

Many other factors influence the vulnerability of a population to disaster. Cutter et al. (2003) were able to reduce 42 variables to 11 independent factors to create a Social Vulnerability Index (SoVI). The Hazards and Vulnerability Research Institute (HVRI) at the University of South Carolina compiled these factors, which include race and class, wealth, elderly residents, Hispanic ethnicity, special needs individuals, Native American ethnicity, and service industry employment, to provide county-level measures for the entire United States.

Figure 9.15 illustrates the use of these data in constructing the state hazard mitigation plan for Mississippi. While the HVRI separates these data into quintiles and then combines the median 3 to create high-, medium-, and

FIGURE 9.14
Density of mobile home housing stock in Mississippi (2010).

low-vulnerability classes, these classes were disaggregated by Mississippi to suit local needs established by the Hazard Mitigation Planning Council. Regardless, SoVI serves functionally as an informative measure of both susceptibility to disaster and resilience after the event has passed.

The final stage of vulnerability analysis involves the establishment of a scoring system by which vulnerability may be assessed against any particular threat. This phase of the mitigation planning process is somewhat subjective in that it involves ranking or scoring different elements of the threat assessment and vulnerability analysis to create a final ranking of threats with respect to their overall risks.

As of this writing, no nationally consistent means for doing so was readily identifiable. The state of Mississippi, recognizing the need for consistency

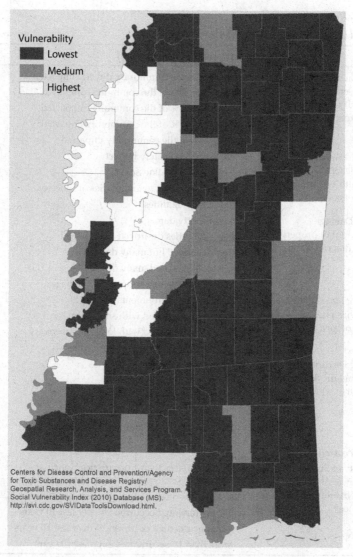

FIGURE 9.15

(See color insert.) Social Vulnerability Index for Mississippi counties (2006).

for the distribution of funds in support of related mitigation activities, developed and adopted the methodology identified in Tables 9.3 and 9.4.

Once the risk characterization was completed for each identified natural hazard, the sum of the risk characteristics was multiplied by their probability of occurrence to determine each hazard's total risk rating score. The maximum score possible is 100. Table 9.5 provides a recap of the risk level attained for each hazard.

TABLE 9.3

Natural Hazard Identification Methodology

Risk Characteristic (Vulnerability)		Score
Area Impacted	No area in the community directly impacted	0
(The percent of the	Less than 25% of the community impacted	1
community at risk to an	Less than 50% of the community impacted	2
impact from each hazard)	Less than 75% of the community impacted	3
	Over 75% of the community impacted	4
Health and Safety	No health and safety impact	0
Consequences	Few injuries or illnesses	2
(The health and safety	Few fatalities but many injuries or illnesses	3
consequences that can occur)	Numerous fatalities	4
Property Damage	No property damage	0
(The amount of property	Few properties destroyed or damaged	1
damage that can occur)	Few destroyed but many damaged	2
	Few damaged and many destroyed	3
	Many properties destroyed and damaged	4
Environmental Damage	Little or no environmental damage	0
(The environmental damage	Resources damaged with short-term recovery	1
that can occur)	Resources damaged with long-term recovery	2
	Resources destroyed beyond recovery	4
Economic Disruption	No economic impact	0
(The economic disruption that	Low direct and indirect costs	1
can occur)	High direct and low indirect costs	2
	Low direct and high indirect costs	3
	High direct and high indirect costs	4
Future Occurrence		
Probability of Future	Unknown/less than 1 occurrence, anticipated rare	1
Occurrence	occurrence	
(The probability of a future	1–4 documented occurrences over last 10 years	2
occurrence)	5–7 documented occurrences over last 10 years	3
	8–10 documented occurrences over last 10 years	4
	More than 10 occurrences over last 10 years	5

9.4 Establishing Capability Targets and Implementing the Plan

Table 9.5 identifies the quantified, documented risk as derived from assets
(CIKR), threats, and vulnerability using both hard data obtained from quan-
tifiable data sets and soft data obtained through the input of information
and experience of the hazard mitigation committee. Theoretically, all assets
could be hardened against any threat and vulnerability completely eliminated in a world with infinite resources.

TABLE 9.4

Natural Hazard Rating Score Schema

Risk Level	Total Rating Score
Low	*0–33*

A hazard with a low risk rating is expected to have little to no impact upon the community. The hazard poses minimal health and safety consequences to the state's residences and is expected to cause little to no property damage. The occurrence of a hazard with a low risk rating is rare; however, due to other factors such as geographical location, it is still possible for such a hazard to occur and even cause significant damage based upon the magnitude of the event.

Medium	*34–67*

A hazard with a medium risk rating is expected to have a moderate impact upon the community. The hazard poses minor health and safety consequences with minor injuries expected and few to no fatalities. The hazard may cause some properties to be damaged or destroyed. The occurrence of a hazard with a medium risk rating is likely at least once within the next 25 years.

High	*68–100*

A hazard with a high risk rating is expected to have a significant impact upon the community. The hazard poses high health and safety consequences with numerous injuries and fatalities possible. The hazard may even cause some properties to be damaged or destroyed. A hazard with a high risk rating is expected to occur at least once within a 12-month period, but can occur multiple times within a year.

Reality dictates otherwise, and the mitigation planning team must seek public input and the guidance of leadership as to how best to allocate limited resources within the context of the THIRA study. Mitigation projects that offer the greatest return on investment with respect to the highest risks should be prioritized over those requiring substantial investment and only a modest incremental increase in overall risk reduction.

This is the final stage of the mitigation planning process and represents the greatest opportunity for creative thought. Thus, the input sought by analysis should be as diverse as possible. Nonsensical ideas may abound, and personalities may at times be difficult to manage among stakeholder groups, but ultimately a wide base of ideas will often produce the greatest success. For example, Table 9.5 consistently ranked tornado as either a high or medium threat for all jurisdictions covered by the MEMA Region 3 plan.

Discussion of the data yielded the following proposals for risk reduction with respect to this threat, which require only minimal investment:

1. Identify the clusters of mobile homes and build community storm shelters rather than individual ones.
2. Create ordinances requiring minimal structural integrity and limited ages for mobile home occupancy permits.

TABLE 9.5

Summary of Hazard Ranking

Hazard Type	Attala	Bolivar	Carroll	Holmes	Humphreys	Leflore	Montgomery	Sunflower	Washington
Dam/levee failure	Low	Low	Medium	Low	Low	Low	Low	Low	Low
Drought	Low	Low	Low	Low	Low	Low	Low	Low	Low
Earthquake	Low	Low	Low	Low	Low	Low	Low	Low	Low
Excessive heat/heat	Low	Low	Low	Low	Low	Low	Low	Low	Low
Expansive soils	Low	Low	Low	Low	Low	Low	Low	Low	Low
Flood	High	High	Low	Medium	High	Medium	Low	High	High
Severe storm	Medium	Medium	Medium	Medium	Medium	Medium	Medium	Medium	High
Severe winter weather	Medium	Medium	Low	Low	Low	Medium	Low	Medium	Medium
Tornado	Medium	Medium	Low	High	Medium	Medium	Low	Medium	Low
Tropical storm/hurricane	Low	Low	Low	Low	Low	Low	Low	Low	Low
Wildfire	Low	Low	Low	Low	Low	Low	Low	Low	Low

3. Use GIS to analyze the effective distance of existing storm warning sirens and use the results to justify future grant applications for new sirens.

4. Establish a public tree trimming program whereby dead trees prone to strong winds would pose significant hazards.

5. Work to align long, aboveground elements of critical infrastructure susceptible to tornado in a southwest to northeast direction (parallel to the tracks of most tornados, rather than potentially perpendicular).

The net effect of applying a sound hazard mitigation planning strategy is the prioritization of resource allocation for reducing risk. The principles and methods presented in this chapter may be applied at almost any geographic scale and with respect to any population, community, or element of critical infrastructure.

Geospatial technologies serve as a highly relevant toolset for creating information to drive data-driven decision making supplemented with local knowledge and experience. Herein lies the greatest application area for GIS and remote sensing, as a life, home, or infrastructure element saved is far more valuable than one lost to a disaster.

References

Brooks, T., and Boone, R. (2014). Mississippi Emergency Management Agency Region 3 hazard mitigation plan, Cleveland, MS.

Buckle, P. (1995). A framework for assessing vulnerability. *Australian Journal of Emergency Management*, 10(1), 11–15.

Cutter, S.L., Boruff, B.J., and Shirley, W.L. (2003). Social vulnerability to environmental hazards. *Social Science Quarterly*, 84(2), 242–261.

Department of Homeland Security. (2013). *Threat and Hazard Identification and Risk Assessment Guide: Comprehensive Preparedness (CPG) 201*. 2nd ed.

Esri. (n.d.). Help file for Moran I tool. Esri, Redlands, CA.

Federal Emergency Management Agency. (2007). Core capability/target capability crosswalk. http://www.fema.gov/media-library/assets/documents/29225.

Federal Emergency Management Agency. (2013). *Local Mitigation Planning Handbook*.

Federal Emergency Management Agency. (n.d.). *Hazus*. http://www.fema.gov/hazus.

Godschalk, D.R. (1991). Disaster mitigation and hazard management. In *Emergency Management: Principles and Practice for Local Government*, ed. T.E. Drabek and G.J. Hoetmer, 131–160. International City Management Association, Washington, DC.

Monmonier, M. (1998). *Cartographies of Danger*. University of Chicago Press, Chicago.

Poisson, S.D. (1837). *Recherches sur la probabilité des jugements en matière criminelle et en matiere civile, precedes des régles generals du calcui des probabilités*. Bachelier, Impremeur-Libraraire Pour Les Mathématiques et Physique, Paris.

10

Conclusion

10.1 Cyber Security

As every computer user is aware, there are a variety of risks associated with computer infection by malicious software or *malware*. Malware is defined by its malicious intention, as opposed to software that is deficient or defective and, as a result, may be unintentionally destructive. Malware takes many forms, including computer viruses, worms, and Trojan horses. It also includes ransom ware, which restricts access to the computer until a fee is paid, and spyware, which gathers information about the computer that is infected (and its user), but which may also investigate other computers connected by networking.

Another type of malware that is of increasing concern to information security professionals is the Low Orbit Ion Cannon (LOIC), which was designed originally to perform network stress testing through a denial-of-service attack. Because the original software was released as open source, it is readily available and has been used to great effect by Anonymous and others to attack the networks serving institutions as diverse as the Church of Scientology and the Recording Industry Association of America, and the websites of companies and organizations that opposed the work of WikiLeaks.

In 2012, the hacker collective known as Anonymous used a LOIC to launch a denial-of-service attack against several entities believed to be responsible for the closure of Megaupload Ltd. The targets included the U.S. Department of Justice, the U.S. Copyright Office, the FBI, and several film and music industry associations.

The most visible malware attacks, while harmful to productivity and potentially destructive, were not designed specifically to attack infrastructure. Certainly, computer security professionals were concerned with the impact of such attacks on water systems, but for the most part, they considered malware an issue apart from critical infrastructure protection. That changed in June 2010 with the discovery of the small (500 kB) Stuxnet worm in Kaspersky Lab (Kushner, 2013). A senior researcher at Kaspersky determined that the worm had infected 14 industrial sites in Iran, including nuclear fuel processing facilities.

The worm was installed on microcomputers in Iran through the insertion of a USB memory stick by unknowing users. Once installed, the worm searched first for Microsoft Windows machines and began to replicate. It then began a search through the local network for evidence of Siemens Step 7 software, which is used widely for certain industrial operations, including control systems. It also searched for programmable logic controllers, allegedly for the purposes of causing fast-spinning centrifuges used in nuclear fuel processing to accelerate and tear apart (Zetter, 2014; GReAT, 2014).

Stuxnet became a catalyst for debate over the future of cyber warfare. Once hidden to preserve customer confidence, the risks associated with cyber attacks were in squarely in the public eye. In 2012, Margaret Warner stated,

> We turn to a new cyber campaign against American banking giants and growing worries about what they might foreshadow. It began late last month and continues to this day. Two more U.S. banks are the latest targets in the spate of cyber-hits on American financial institutions. This week, Capital One and BB&T suffered disruptions on their websites, leaving customers without access to their accounts. A group calling itself the Qassam Cyber Fighters claimed responsibility and said the attacks are retaliation for an anti-Muslim video. But some U.S. officials, like Connecticut Senator Joe Lieberman, blame the recent uptick of attacks on Iran and its elite security force. (PBS Newshour, 2012)

In response to Ms. Warner's comment, Senator Joe Lieberman said, "I think that this was done by Iran and the Quds Force, which has its own developing cyber-attack capacity, and I believe it was a response to the increasingly strong economic sanctions." Iran denied any role in the disruption. In any event, Secretary of Defense Leon Panetta stated, "The collective result of these kinds of attacks could be a cyber Pearl Harbor, an attack that would cause physical destruction and the loss of life" (PBS Newshour, 2012).

In 2013, just months after Secretary Panetta's comments, President Obama issued Executive Order 13636, "Improving Critical Infrastructure Cybersecurity" (Obama, 2013). Section 4 of that order states,

> Cybersecurity Information Sharing. (a) It is the policy of the United States Government to increase the volume, timeliness, and quality of cyber threat information shared with U.S. private sector entities so that these entities may better protect and defend themselves against cyber threats. Within 120 days of the date of this order, the Attorney General, the Secretary of Homeland Security (the "Secretary"), and the Director of National Intelligence shall each issue instructions consistent with their authorities and with the requirements of section 12(c) of this order to ensure the timely production of unclassified reports of cyber threats to the U.S. homeland that identify a specific targeted

entity. The instructions shall address the need to protect intelligence and law enforcement sources, methods, operations, and investigations. (Obama, 2013)

Among several responses to this order, the National Institute of Standards and Technology (NIST) published the report *Framework for Improving Critical Infrastructure Cybersecurity* (NIST, 2014). The NIST confirmed its understanding of the nature of critical infrastructure:

Critical infrastructure is defined in the EO as "systems and assets, whether physical or virtual, so vital to the United States that the incapacity or destruction of such systems and assets would have a debilitating impact on security, national economic security, national public health or safety, or any combination of those matters." (NIST, 2014, p. 3)

Henry Kissinger wrote:

Internet technology has outstripped strategy or doctrine—at least for the time being. In the new era, capabilities exist for which there is as yet no common interpretation—or even understanding.... When individuals of ambiguous affiliation are capable of undertaking actions of increasing ambition and intrusiveness, the very definition of state authority may turn ambiguous. The complexity is compounded by the fact that it is easier to mount cyberattacks than to defend against them, possibly encouraging an offensive bias in the construction of new capabilities.

The danger is compounded by the plausible deniability of those suspected of such actions and by the lack of international agreements for which, even if reached, there is no present system of enforcement.... Electric grids could be surged and power plants disabled through actions undertaken exclusively outside a nation's physical territory (or at least its territory as traditionally conceived). (Kissinger, 2014; Kindle edition, n.d.)

Having confirmed that a new danger exists for critical infrastructure, is there a role for geographic information systems (GIS) in mitigating or removing this risk? While technology continues to evolve, at least based on examples, the benefit of GIS for cyber security seems clear (Abdulla, 2004; Wolthusen, 2005).

In its entry for Low Orbit Ion Cannon, Wikipedia notes, "LOIC attacks are easily identified in system logs, and the attack can be tracked down to the IP addresses used at the attack" (Wikipedia, 2014). The British Broadcasting Corporation uses Geo-IP addressing technology *preventively* to restrict access to certain programming. GIS platforms are well suited to support detailed planning for the use of Geo-IP technology. Indeed, during the 2012 Republican

National Convention (RNC), the city of Tampa used a similar approach to block selected traffic coming from geographically specific attackers.

10.2 Natural and Man-Made Disasters

The Old Testament book of Genesis tells the story of Noah and the ark he built. Before Genesis was written, Gilgamesh recounted in his *Epic* the story of Utnapishtim, survivor of a great flood. Like Noah, he had been warned (by the water god Enki) to build a great ship to save his family and animals. Many other cultures throughout the world have similar myths, including legends regarding the coastline of Antarctica.

These similarities have prompted many scholars to investigate the origin of these great flood legends (or histories). The last ice age reached its greatest extent approximately 22,000 years ago and ended about 12,000 years ago. Since that time, there have been at least three great periods of ice melt that have raised sea levels worldwide, at least one of which has been theorized to be the source of the great flood legends.

The archeological record is unclear. We cannot, based on extant evidence, prove that all life on this planet, except one boatload of people and animals, died in a flood. Certainly, absence of evidence is not evidence of absence. However, we can say with confidence that the coastal flooding associated with a 100 m rise in mean sea level would kill large numbers of people (National Geographic Society, 1999).

In 1780, a hurricane passed through the Caribbean en route to the southeastern United States. Know as the Great Hurricane of 1780 or Huracán San Calixto, the storm caused the death of as many as 22,000 people. Winds in excess of 320 km/h were inferred from the effects of the storm: in Barbados, large-bore cannons were thrown 30 m into the air and the bark was stripped from trees. In terms of human life, this was the most costly hurricane of which we have confirmation.

Blizzards and monsoon rains, volcanoes and earthquakes, tsunamis and tidal waves, hurricanes and cyclones, avalanches and mud—all have taken their toll on human life. Studied for millennia, these natural disasters are coming, albeit slowly, to be understood. GIS has become an integral part of that effort.

For example, data concerning plate tectonics and continental drift, combined with field observations of floods recorded over centuries, have been synthesized using geographic analysis in general and GIS in particular to identify areas most prone to tsunamis. Knowing the correct general location has allowed scientists to place, more efficiently and effectively, floating detectors of subsurface motions to provide advance warning to residents of potentially affected coastal areas.

By itself, GIS cannot explain why sea levels are rising. However, it can be used with elevation data derived from digital elevation models (DEMs) to predict the impact of those rising sea levels. While debate continues about the root cause of the problem, tangible actions can be taken to preserve life and property using GIS.

On September 11, 2001, the world changed (Cutter et al., 2003). Some have argued that the date was selected to commemorate the Battle of Vienna, which took place on September 11, 1683, and which effectively ended the expansion of the Turkish Ottoman Empire into Western Europe. Others have argued that the date was selected simply as a matter of convenience: planning was complete; it was a Tuesday, so airline traffic would be light; Congress would be in session; and the office buildings would be full. Regardless of the actual reason, the coincidence of the date—9/11—and the use of the telephone number 911 for emergency calls in the United States was noted immediately and would serve as a constant reminder of the day's villainy.

It is not an overstatement to say that everything changed, not just in the United States, but throughout the world. One of the authors (Austin) traveled to China 3 weeks later. Many colleagues expressed concern about the safety of travel to a foreign country. Upon landing in Hong Kong, the first news that appeared in the local press was the movement of 350,000 Chinese troops to the border with the Xinjiang Uyghur Autonomous Region to manage any collateral efforts by similarly radicalized individuals.

The responses to 9/11 have affected lives throughout the world. Most obvious, perhaps, was the U.S. invasion of Iraq on March 20, 2003. It may be argued that the overthrow (in 2003) and execution (in 2006) of Saddam Hussein Abd al-Majid al-Tikriti in Iraq served as a catalyst for other revolutions. Indeed, some argue that this was a direct precursor to the Arab Spring uprisings that began in December 2010, to the fall of Muhammad Hosni El Sayed Mubarak in Egypt and Muammar Muhammad Abu Minyar al-Gaddafi in Libya in 2011, and to the Syrian civil war that began in 2011. In a sense, the evolution of the Islamic State of Iraq and the Levant into the Islamic State brings closure and a return to the events of September 11, 1683.

The Geneva Conventions consist of four treaties, signed in 1864, 1906, 1929, and 1949.

> The Geneva Conventions and their Additional Protocols form the core of international humanitarian law, which regulates the conduct of armed conflict and seeks to limit its effects. They protect people not taking part in hostilities and those who are no longer doing so. (International Committee of the Red Cross, 2014)

The Geneva Conventions have been signed by 196 nations. The war against terrorists is not a war between nations, regardless of the claims made to statehood by the Islamic State of Iraq and the Levant. Rather, it is a war between cultures and ideologies and, as such, is not subject to the constraints of the Geneva Conventions.

The National Commission on Terrorist Attacks upon the United States (2014b), also known as the 9/11 Commission, provided a detailed (585-page) analysis of what happened on September 11, 2001. As the commission wrote in the executive summary to the report, "The nation was unprepared" (National Commission on Terrorist Attacks upon the United States, 2014a, p. 1). As documented in this text, the nation has to work diligently in the coming years to adapt our understanding of natural disaster response to the special circumstances of responding to man-made disasters and terrorism.

10.3 Conclusion

The overall goal of this book is to raise awareness of how GIS-based technologies can be used to support critical infrastructure protection and emergency management efforts and to gain knowledge of gaps in preparedness planning, response, recovery, and emergency management activities involving large-scale disasters. In the case of the Geospatially Enabling Community Collaboration (GECCo) exercises, this was accomplished by exploring response and recovery activities in a prolonged and cascading series of disruptions exacerbated by extensive regional physical and cyber infrastructure interdependencies with major complicating factors that exceeded a community's contingency planning and response capabilities of most critical infrastructures and essential emergency service providers.

Several hundred observations, findings, lessons learned, and recommendations are identified throughout this text. They are based on the experiences of the authors and on the experiences of countless representatives from local, regional, state, and federal agencies, from the emergency management community, from infrastructure companies, and from private organizations.

On critical infrastructure interdependencies, the findings and recommendations focused on the need to develop a common set of assumptions with common terminology for worst-case scenarios to provide organizations with a common baseline for risk assessments and exercises. There was common agreement about the need to make organizations aware of the importance of incorporating interdependencies into vulnerability and risk assessments.

A major focus area involved a combination of transportation, municipal utility, and energy interdependencies and the need for conducting regular interdependency exercises to better understand the cascading effects of critical infrastructure. It was often noted that there was a formal structure for coordinating the activities of emergency planning and protection of infrastructure between the federal and local government agencies and public utility companies.

On the topic of situational awareness, GECCo participants discussed how emergency management and infrastructure owners communicate with each other during emergencies to find out what is down and how to respond. It was often noted that that public and private sector organizations need to decide collectively on reservation of frequencies so needs are met, and procedures need to be developed to allow organizations and individuals to use these systems during an emergency.

In the case of emergency communications and associated IT infrastructure resiliency, it was consistently noted that these systems are a priority for government at all levels, but not as much in the private sector. The ongoing need to improve interoperability among these systems was also identified as vital to emergency response and management during a disaster.

As for information sharing and collaboration, GECCo participants wholly agreed that it is critical to set up prearranged data-sharing agreements, protocols, and standards, and to ensure inclusion of all key local, regional, state, and federal stakeholders, including private sector organizations.

Most local government agencies, utilities, and telecommunications representatives indicated that private companies are reluctant to share information directly with government organizations. Many utilities noted that a trust relationship is paramount in creating an environment where it is felt that information can be shared safely, and in confidence (Jones, 2005).

Regarding contingency planning and mitigation measures, there was concern consistently expressed that no centralized point exists to collect information and share best practices, and that there are many good ideas and resources that exist that can provide significant benefits if there is awareness of what these are, and points of contact that can provide information on how to leverage these capabilities. It was often suggested that an online portal could be developed as a collaborative mechanism at the local, regional, state, and tribal levels.

Another major area involved resilient communications. Among the various GECCo events, a majority of issues centered on the question of how to determine risk and identify necessary mitigation activities. In this context, the participants discussed how to determine what emergency communications contingency plans and capabilities needed developed and how to determine which organizations should be involved.

While the government agencies and the private sector have performed a variety of risk assessments on their own operations and systems, they have not necessarily recognized the dependencies with other organizations that they are reliant on during times of a disaster, and that these interdependencies are fully known. For this reason, risk assessment and mitigation need to be performed on a regional basis, and larger organizations need to share the resources to undertake these measures because smaller entities lack the capability.

Several points were made throughout this text regarding best practices. Some of these best practices were specific to environmental circumstances (for example, hurricanes in the coastal southeastern United States). Some were specific to political circumstances (for example, national political party conventions). In some instances, the two circumstances became inextricably entwined.

The best practices, recommendations, and key learning points can be boiled down to a few simple principles, which serve as our conclusion.

Planning. Plan your response as far in advance of the event as possible. After you finish your plan, revise it often to reflect changes in local circumstances, changes in technology, and changes in the external support environment. Manage these plans on a GIS platform to allow users to become familiar with the tools and capabilities of GIS *before* the event.

Collaboration. As part of the planning process, identify appropriate agencies, organizations, and individuals who can work beneficially with you during an event response. Prepare and sign memoranda of understanding or other appropriate documents to enable interaction during the event.

Data sharing. Share data with your collaboration partners before the event. At the very least, share data structures and metadata to permit expedited data sharing in your GIS environments during the event.

Communication. Let users and the public know of your capabilities. Make provisions for public demonstrations to encourage support from and participation by the local community.

References

Abdalla, R. (2004). Utilizing 3D web-based GIS for infrastructure protection and emergency preparedness. In *XXth ISPRS Conference Proceedings, Technical Commission VII*, vol. XXXV, part B7, pp. 653–657.

Cutter, S.L., Richardson, D.B., and Wilbanks, T.J., eds. (2003). *The Geographical Dimensions of Terrorism*. Routledge, New York.

GReAT. (2014). Stuxnet: Zero victims—The identity of the companies targeted by the first known cyber-weapon. *SecureList*, November 11. http://securelist.com/analysis/publications/67483/stuxnet-zero-victims/.

International Committee of the Red Cross (ICRC). (2014). Geneva Conventions. https://www.icrc.org/en/war-and-law/treaties-customary-law/geneva-conventions (accessed December 17, 2014).

Jones, B.A. (2005). Identifying sensitive critical infrastructure data. Presented at the Geospatial Information and Technology Association Conference Proceedings, Denver.

Kissinger, H. (2014). *World Order*. Penguin Press, New York (Kindle edition).

Kushner, D. (2013). The real story of Stuxnet: How Kaspersky Lab tracked down the malware that stymied Iran's nuclear-fuel enrichment program. *IEEE Spectrum.* http://spectrum.ieee.org/telecom/security/the-real-story-of-stuxnet.

National Commission on Terrorist Attacks upon the United States. (2004a). *The 9/11 Commission Report Executive Summary.* http://govinfo.library.unt.edu/911/report/911Report_Exec.pdf.

National Commission on Terrorist Attacks upon the United States. (2004b). *The 9/11 Commission Report.* http://govinfo.library.unt.edu/911/report/index.htm.

National Geographic Society. (1999). Ballard and the search for the Black Sea. http://www.nationalgeographic.com/blacksea/ax/frame.html.

National Institute of Standards and Technology. (2014). *Framework for Improving Critical Infrastructure Cybersecurity.* Version 1. February 12. http://www.nist.gov/cyberframework/upload/cybersecurity-framework-021214.pdf.

Obama, B. (2013). Improving critical infrastructure cybersecurity. Executive Order 13636, DCPD-201300091. February 19. http://www.gpo.gov/fdsys/pkg/FR-2013-02-19/pdf/2013-03915.pdf.

PBS Newshour. (2012). Could the U.S. face 'cyber Pearl Harbor'? Protecting banks from hacker attacks. October 18. http://www.pbs.org/newshour/bb/science-july-dec12-cyber_10-18/.

Rich, S., and Davis, K.H. (2010). Geographic information systems (GIS) for facility management. IFMA Foundation, Houston, TX. http://foundation.ifma.org/docs/default-source/Whitepapers/foundation-geographic-information-systems-%28gis%29-technology.pdf?sfvrsn=2.

U.S. Department of Homeland Security. (2004). *Progress and Challenges in Securing the Nation's Cybersecurity.* OIG-04-29. Office of the General Inspector, Washington, DC.

Wikipedia. (2014). Low Orbit Ion Cannon. http://en.wikipedia.org/wiki/Low_Orbit_Ion_Cannon (accessed December 16, 2014).

Wolthusen, S.D. (2005). GIS-based command and control infrastructure for critical infrastructure protection. Presented at the First IEEE International Workshop on Critical Infrastructure Protection, Darmstadt, Germany.

Zetter, K. (2014). *Countdown to Zero Day: Stuxnet and the Launch of the World's First Digital Weapon.* Crown Publishing, New York.

Index

Printed in the United States
by Baker & Taylor Publisher Services